Assessing Revolutionary and Insurgent Strategies

SPECIAL TOPICS IN IRREGULAR WARFARE:
UNDERSTANDING RESISTANCE

Erin N. Hahn, Editor

Katharine Raley Burnett, David J. Danelo, Eric Dunford,
James A. Gavrilis, Erin N. Hahn, Jameel Khan, Jesse Kirkpatrick,
W. Sam Lauber, Margaret McWeeney, Summer Newton,
Guillermo Pinczuk, Theodore Plettner, Joseph M. Tonon, and
Timothy Wittig, Contributing Authors

United States Army Special Operations Command
and
The Johns Hopkins University Applied Physics Laboratory
National Security Analysis Department

ASSESSING REVOLUTIONARY AND INSURGENT STRATEGIES

The Assessing Revolutionary and Insurgent Strategies (ARIS) series consists of a set of case studies and research conducted for the US Army Special Operations Command by the National Security Analysis Department of the Johns Hopkins University Applied Physics Laboratory.

The purpose of the ARIS series is to produce a collection of academically rigorous yet operationally relevant research materials to develop and illustrate a common understanding of insurgency and revolution. This research, intended to form a bedrock body of knowledge for members of the Special Forces, will allow users to distill vast amounts of material from a wide array of campaigns and extract relevant lessons, thereby enabling the development of future doctrine, professional education, and training.

From its inception, ARIS has been focused on exploring historical and current revolutions and insurgencies for the purpose of identifying emerging trends in operational designs and patterns. ARIS encompasses research and studies on the general characteristics of revolutionary movements and insurgencies and examines unique adaptations by specific organizations or groups to overcome various environmental and contextual challenges.

The ARIS series follows in the tradition of research conducted by the Special Operations Research Office (SORO) of American University in the 1950s and 1960s, by adding new research to that body of work and in several instances releasing updated editions of original SORO studies.

VOLUMES IN THE ARIS SERIES

Casebook on Insurgency and Revolutionary Warfare, Volume I: 1927–1962 (Rev. Ed.)
Casebook on Insurgency and Revolutionary Warfare, Volume II: 1962–2009
Case Studies in Insurgency and Revolutionary Warfare—Colombia (1964–2009)
Case Studies in Insurgency and Revolutionary Warfare: Cuba 1953–1959 (pub. 1963)
Case Study in Guerrilla War: Greece During World War II (pub. 1961)
Case Studies in Insurgency and Revolutionary Warfare: Guatemala 1944–1954 (pub. 1964)
Case Studies in Insurgency and Revolutionary Warfare—Palestine Series
Case Studies in Insurgency and Revolutionary Warfare—Sri Lanka (1976–2009)
Unconventional Warfare Case Study: The Relationship between Iran and Lebanese Hizbollah
Unconventional Warfare Case Study: The Rhodesian Insurgency and the Role of External Support: 1961–1979
Human Factors Considerations of Undergrounds in Insurgencies (2nd Ed.)
Irregular Warfare Annotated Bibliography
Legal Implications of the Status of Persons in Resistance
Narratives and Competing Messages
Special Topics in Irregular Warfare: Understanding Resistance
Threshold of Violence
Undergrounds in Insurgent, Revolutionary, and Resistance Warfare (2nd Ed.)

SORO STUDIES

Case Studies in Insurgency and Revolutionary Warfare: Vietnam 1941–1954 (pub. 1964)

TABLE OF CONTENTS

ILLUSTRATION CREDITS

The authors acknowledge the sources of illustrations included in this study.

INTEREST, IDENTIFICATION, INDOCTRINATION, AND MOBILIZA-TION: I³M. Figures 1 and 2 first published in Jesse Kirkpatrick and Mary Kate Schneider, "I³M–Interest, Identification, Indoctrination, and Mobilization: A Short Introduction to a New Model of Insurgent Involvement," *Special Warfare Magazine* 26, no. 4 (October–December 2013): 23–27.

THRESHOLD OF VIOLENCE. Figures 2 and 3 first published in Guillermo Pinczuk, Mike Deane, and Jesse Kirkpatrick, *Case Studies in Insurgency and Revolutionary Warfare—Sri Lanka (1976–2009)*, ed. Guillermo Pinczuk (Fort Bragg, NC: US Special Operations Command, 2014).

THREAT FINANCE. Table 1 first published in *Vulnerability Assessment Method: A Practitioner's Handbook* (Fort Meade, MD: United States Army Asymmetric Warfare Group, 2010).

PREFACE

This Special Topics volume is an introduction to subjects related to the broad concept of resistance. Over the past six years, Assessing Revolutionary and Insurgent Strategies (ARIS) researchers have used the term *resistance* to cover a spectrum of actors who challenge existing social structures. Their research highlights the complex and dynamic characteristics of resistance and its associated concepts. The ARIS project has largely focused on resistance movements that emerge against a government or political actor, although the work acknowledges the very broad range of structures a resistance may confront (e.g., social norms and nongovernmental institutions or organizations). The research seeks to understand how and why a resistance develops and aims to identify emerging trends in operational designs and patterns relevant to special operations missions.

In pursuit of that goal, this volume covers topics that were identified as important to improving our understanding of fundamental issues associated with resistance movements. It provides an introduction to the relevant concepts, and in some cases, more in-depth studies will be undertaken. It covers wide-ranging topics, such as how a resistance organizes and develops ("Exploring the Phases of Contemporary Resistance" and "Interest, Identification, Indoctrination, and Mobilization: I³M") as well as discrete elements of a movement's organizational structure and methods ("The Public Component" and "Threshold of Violence"). It also explores concepts relevant to a resistance movement's ability to eventually govern ("How Resistance Groups Generate Legitimacy" and "Transition from Resistance to Governance"), among other subjects.

The ideas introduced in this volume should be seen as starting points for further analysis—a way of presenting special operations personnel and others with a set of concepts that are important to consider and grapple with so that they can gain a more complete assessment of resistance and its complexity. This, like the other ARIS products, is a learning tool that highlights just some of the elemental structures and attributes of resistance. After studying these topics, special operations personnel can apply their knowledge professionally.

EXPLORING THE PHASES OF CONTEMPORARY RESISTANCE

David J. Danelo

Political resistance, like war, is both timeless and ever-changing. Although twenty-first-century resistance organizations form and disperse by using increasingly varied methods and means, they do so along a continuum of phases that is similar to stages of warfare in past eras. These phases and stages are important for scholars to study—and for soldiers to understand—for two reasons. First, students need to achieve a more accurate understanding of what modern unconventional warfare (UW) and counterinsurgency (COIN) look like. Second, and more importantly, these studies equip soldiers (and their political leaders) with essential knowledge that will allow them to fight and win irregular conflicts by understanding how resistance movements move through each phase and the implications of each phase on operational planning.

Why is understanding the phases of contemporary resistance important? Understanding the sequential phases of organizational development in contemporary resistance, alongside the determinants and variables that compose each phase, enables Special Operations Forces planners to determine the course of action most likely to either enhance resistance or disrupt insurgent activity. Additionally, developing and understanding a phasing construct will assist in mission planning, ultimately leading to more effective adaptation and fluid decision cycles while operators are deployed. Although the phases of resistance are often fluid, each phase is distinct and requires appropriate planning and tools. Deeper analysis and study of the phasing of contemporary resistance movements will result in application of the equipment, gear, and tactics appropriate for the particular environment.

The purpose of this paper is to educate readers on the phases of contemporary resistance and propose a construct for more detailed research, study, and analysis. The method of study the authors followed when developing this document included deconstruction of existing literature, explanation of existing determinants and variables within current UW and insurgency phasing constructs, tabulated analysis of variables related to the phases, and recommendation for further study. After reading this paper, readers should have confidence in their understanding of existing literature and doctrine on resistance phasing; a thorough appreciation for the complexity of determinants and variables that influence changes in phases; and a detailed understanding of the additional research required to analyze the unknown variables and develop the proposed phasing construct.

DEFINING RESISTANCE

Separate definitions for resistance and insurgency exist within the Department of Defense and various academic communities. Generally speaking, the term *insurgency* describes a movement designed to achieve political goals through armed conflict against an indigenous government or occupying power. *Resistance* suggests an insurgency focused on removing an occupying power. UW, according to US military doctrine, involves enabling, supporting, facilitating, or advising resistance movements to further US political objectives. COIN involves halting or reversing resistance movements. Doctrine describing both UW and COIN is relevant for understanding and analyzing the phases of contemporary resistance.

The phases of contemporary resistance also frame the narrative arc of resistance and insurgent groups. The use of narrative in resistance is not a new phenomenon, and during the American Revolution, colonial patriots slanted news in their favor, proselytized American national identity, and shaped their story to justify guerrilla tactics. Isaias Afewerki struggled for three decades to achieve Eritrean independence from Ethiopia, finally turning the last page in his own heroic struggle in 1991 when he became Eritrea's first (and only) president. In a more recent example, the Sunni insurgency that has evolved into Islamic State envisions the final, absolute end of its struggle as a caliphate dominating the entire world.

UW doctrine also builds a narrative that includes seven phases of resistance that can be tailored to include only those phases required by the political and military situation on the ground. In contrast, Mao Zedong's story about the phases of insurgency is normally framed as a three-act play—although, as we will see, this construct may limit the analysis and ultimately be inadequate for studying the phases of contemporary resistance. Seeing each phase of resistance as a discrete component—linked to but distinct from the other phases—enables Special Operations Forces to design strategies that will either enhance or interrupt the narrative and create an opportunity to write a different ending to the story.

DECONSTRUCTING THE LITERATURE ON PHASES

Mao

No conversation about doctrine on phases of contemporary resistance should start without a discussion of Mao Zedong, the Chinese Communist revolutionary whose twentieth-century theories of insurgent warfare helped him defeat his rivals and inspired parallel movements in North Korea, Cambodia, and Vietnam. Political science students commonly note Mao's three phases of insurgency (and the Internet is replete with references to Mao's three phases), and every US

Army and Marine Corps publication on UW, resistance, insurgency, or COIN references this phasing construct.

Mao's three phases can be summarized as follows:

1. Organization, consolidation, and preservation of base areas, usually in difficult and isolated terrain
2. Progressive expansion by terror and attacks on isolated enemy units to obtain arms, supplies, and political support
3. Decision, or destruction of the enemy in battle[1]

Field Manual 3-24: Mao Made Modern

Although scholars have challenged these three phases of insurgency, the basic construct has endured long enough to become enshrined in current military doctrine. Field Manual (FM) 3-24, *Counterinsurgency*,[2] states the phases of insurgency as follows:

1. Latent and incipient
2. Guerrilla warfare
3. War of movement

In this way, current military doctrine—particularly with respect to countering insurgency—elevates Mao's three phases to an almost sacrosanct status.

Before we set it aside, we should note that the three-phase construct has several benefits. It is simple. It is (or, at least it was) relatively accurate. It is easy to condense, classify, explain, and understand. And the simplicity of the three phases means they can be easily shaped into quotable congressional testimony and cable news sound bites.

Unfortunately, although the three-phase construct simplifies the flow of Mao's narrative, it does not sufficiently explain how Mao operationalized his revolutionary insurgent campaigns. The three phases are quite broad and do not provide a level of detail that is operationally useful. This is the core challenge with any phasing construct: how can the theory be applied to force generation design? How do planners and operators identify which phase an insurgency is in and then create psychological and practical strategies to influence the insurgent campaign? These questions become more challenging to answer when looking at Mao's theory through only the three-phase paradigm.

Mao Reconsidered

Mao himself actually had a more complex view of the phasing process. The following excerpt is from Fleet Marine Force Reference

Publication 12-18, *Mao Tse-tung on Guerrilla Warfare*,[3] a 1961 translation by US Marine Corps Brigadier General Samuel B. Griffith, who, two years later, completed what was then called the definitive translation of Sun Tzu's *The Art of War*.[4]

Consider how Mao describes the campaign sequence in light of the three-phase construct:

> This policy we pursue in order to gain our political goal, which is the complete emancipation of the Chinese people. There are certain fundamental steps necessary in the realization of this policy, to wit:
>
> 1. Arousing and organizing the people
> 2. Achieving internal unification politically
> 3. Establishing bases
> 4. Equipping forces
> 5. Recovering national strength
> 6. Destroying enemy's national strength
> 7. Regaining lost territories
>
> *There is no reason to consider guerrilla warfare separately from national policy.* (Emphasis added.)

If, as Mao states, guerrilla warfare should not be considered separately from national policy—and if the steps proposed to achieve success in contemporary resistance look anything like the ones outlined above—then accomplishing these objectives will require the use of measurable variables to make clear which steps have been completed. Mao's examination of what determines advancement to the next phase was adequate for his purposes—particularly given China's terrain, military resources, and substantial rural population. However, it is difficult to adapt in a contemporary construct because contemporary conflict often emerges in urban settings, features different methods to mobilize the resistance, and may be fought for different reasons (grievances). The differences in contemporary conflict do not mean Mao's phasing is wholly inapplicable, but they do support the need to adapt the phasing construct to account for these differences.

Army Doctrine and Training Publication 3-05: UW and Mao

US Army Doctrine and Training Publication (ATP) 3-05, *Unconventional Warfare*,[5] also describes a seven-phase construct. Although developed as an operational rather than a strategic model, the ATP 3-05 construct appears to mirror Mao's model (the seven phases cited in

Griffith's translation, rather than the simplified three phases outlined in FM 3-24) more than any other source. For this reason, we will consider the seven-phase ATP 3-05 UW phase model below (with Mao's corresponding phase noted parenthetically):

1. Preparation (arousing and organizing the people)
2. Initial contact (achieving internal unification politically)
3. Infiltration (establishing bases)
4. Organization (equipping forces)
5. Buildup (recovering national strength)
6. Employment (destroying enemy's national strength)
7. Transition (regaining lost territories)

Unlike many of the doctrinal publications on COIN, which tend to reflect the bias of having been developed during the middle of a conflict, ATP 3-05's UW approach stands out in the canon on resistance phases for three reasons. First, as mentioned, it parallel's Mao's seven phases. Second, it is easy to operationalize. Third, it is scalable in reverse: UW operators could use the model to understand (and interrupt) an adversary's resistance movement just as easily as they could use the model to build capacity for resistance.

The singular challenge with the ATP UW phase model is its inability to enable identification of determinants and variables that correspond to each phase and to correlate those determinants with stages of organizational growth. Although straightforward and useful for operational planning, the ATP phase model still has weaknesses.

Galula: Mao Meets the Bourgeois-Nationalist "Shortcut"

David Galula's theoretical classic, *Counterinsurgency Warfare: Theory and Practice*,[6] offers another phasing construct. Galula wrote from the perspective of the French-Algerian War, and his work was as defined by that experience as Mao's was by his own revolution. Galula's proposed phasing construct is similar to Mao's; however, he offers a "bourgeois-nationalist shortcut" through the guerrilla warfare pattern that entails waging successive terrorist campaigns. At once both respectful and dismissive, Galula describes Mao's phasing construct as the "orthodox Communist" pattern and compares it to the "shortcut" construct that he saw at work in Algiers. Galula devotes almost a quarter of the book to an eight-step phasing construct for COIN operations, which was also drawn from the French campaign. Forms of this construct have often been operationalized in successive campaigns and are frequently referenced when creating new COIN doctrine (see FM 3-24).

However, there are three problems with Galula's approach. First, contemporary resistance cannot easily be classified as Algeria 2.0; the problem sets that operators and planners contend with can rarely be restructured into the eight-step campaign Galula proposes. In both Iraq and Afghanistan, US and North Atlantic Treaty Organization (NATO) forces conducted four or five steps simultaneously, and the operational designs—and enemy counterattacks—rarely incorporated themselves into the Galula process. They were not carried out in a linear, methodical fashion as Galula believed the insurgent process should progress. Second, as with Mao, Galula is thin on determinants; his work tells operators what to do (e.g., "The Fifth Step: Local Elections" and "The Sixth Step: Testing the Local Leaders") but does not describe the variables at work in reaching those steps. Third—and most often forgotten in the enthusiasm for Galula's theoretical construct—history reminds us that the French did not succeed in defeating the Algerian insurgency. Galula's work is important, and contributes well to the phasing analysis, but it should not be the final voice in developing the problem set.

Special Operations Research Office: The Determinants and Variables Solution

From Mao to FM 3-24 (which is essentially an updated version of Mao) to ATP 3-05 to Galula, we remain challenged with finding the determinants and variables that frame the phase paradigm. This brings us to a significant, and underappreciated, 1966 publication, *Human Factors Considerations of Undergrounds in Insurgencies,* from the Special Operations Research Office (SORO).[7] The SORO publication proposes models of human behavior through multiple phases of resistance. The catalog of factors compiled in the SORO document, as well as the narrative model of five phases proposed early in the publication, offer a more detailed list of determinants and variables than was offered by either Mao or Galula, suggesting a bridge between ATP 3-05 and COIN doctrine. Although there may be better doctrinal prospects for defining the variables that drive insurgency phases, the thoroughness and clarity of the SORO publication are difficult to beat. For this reason, this paper assumes the SORO publication's list of determinants and variables, which are outlined in its table of contents, as a point of departure for analysis. Although not comprehensive, this list offers a noteworthy possibility for framing a more detailed phasing study. Following is a condensed list, derived from the publication's table of contents:

- Underground organization within insurgency
- Organization
- Motivation

- Ideology and group behavior
- Clandestine and covert behavior
- Recruitment
- Education and training
- Finance
- Propaganda and agitation
- Passive resistance
- Terrorism
- Subversive manipulation of crowds
- Planning of missions
- Operations
- COIN intelligence
- Defection programs
- Population control
- Civic action

As a starting point, this paper will take two initial steps when analyzing Mao's phases of insurgency to help widen the aperture for study. First, instead of examining Mao's phasing construct through the usual phases, the paper will use for analysis instead the seven steps as defined in the Griffith translation. Second, to establish a basis for comparison in Mao (and ATP 3-05, as well as any other proposed constructs), this paper will assume the SORO publication's list of determinants and variables as a credible starting point. By deconstructing Mao (and, as a consequence, Galula) into more phases, we will gain a greater understanding of their nuances relative to the model in the SORO publication; examine the utility of the ATP 3-05 and FM 3-24 constructs relative to Mao; and develop a more credible sense of where the phasing paradigm is sufficient when compared with actual experience.

Validating any construct in follow-on studies will require analyzing only two case studies by using a proper methodology; of course, more case studies could be examined. For now, we can take what we know from observation and experience—both from history and from today's headlines—to evaluate the utility of the current phases when measuring the determinants and variables (on one axis of the tables at the end of this paper) against the various phasing constructs (on the other axis of the tables). As with the components of the axes themselves, this serves as a point of departure for comparison, not a definitive conclusion.

The appended tables were developed by extrapolating the variables emphasized in the doctrine or from examples of warfare from the era during which the doctrine was developed. The tables do not provide answers; instead, they illustrate the prospective constructs. To validate the models—and to identify which is superior—the particular details in each table must be developed through proper case studies. Two things matter for our purposes: first, these tables should be accurate representations of how the existing phasing models work, and second, they should serve as a useful methodology for future case studies

Existing Phasing Constructs

Mao's Guerrilla Warfare

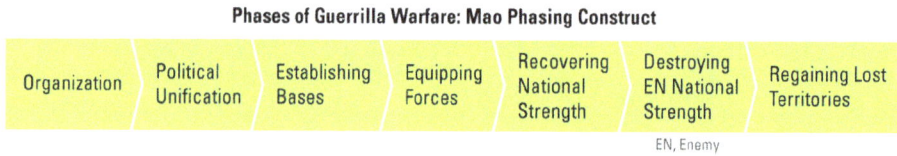

Phases of Guerrilla Warfare: Mao Phasing Construct

Organization	Political Unification	Establishing Bases	Equipping Forces	Recovering National Strength	Destroying EN National Strength	Regaining Lost Territories

EN, Enemy

The effective phasing constructs emerge more clearly by examining each phase and weighing the importance of each determinant and variable within that specific phase. Expanding Mao's construct into seven phases enables a more specific awareness of which determinants have the greatest impact within each phase. This knowledge, in turn, enables operators to more accurately identify which phase an insurgency is in and to produce a more effective solution to counter or enhance the resistance.

ATP 3-05 UW

Phases of UW: ATP 3-05 Unconventional Warfare

Preparation	Initial Contact	Infiltration	Organization	Buildup	Employment	Transition

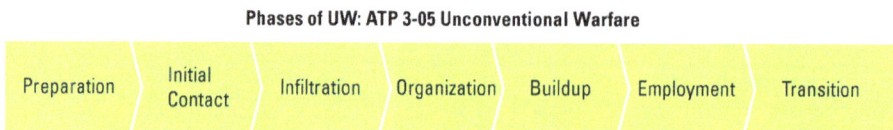

As mentioned, the ATP 3-05 phasing construct bears the greatest similarity to Mao's, staying true to the model without risking oversimplification. Although the model was designed for employing UW, it could potentially be as effective in understanding how to interrupt resistance or insurgency movements, provided researchers could reach a better understanding of how the determinants and variables influenced each phase and how organizations changed through the phases.

Galula's Bourgeois-Nationalist

Phases of Insurgency: Galula Bourgeois-Nationalist "Shortcut" Phasing Construct

Organization	Blind Terrorism	Selective Terrorism	Establishing Bases	Equipping Forces	Recovering National Strength	Destroying EN National Strength	Regaining Lost Territories

↳ Replaces "Political Unification" in Mao Construct   EN, Enemy

As noted before, the Galula model differs little from Mao's model; the substantive difference is the substitution of blind terrorism and selective terrorism (Galula's phases 2 and 3, respectively) for political unification (Mao's phase 2). Although this construct may accurately model Galula's perspective against Mao's, more importantly, it gives planners and operators a greater likelihood of accurately identifying which phase an insurgency is in. Galula believed insurgency was carried out step by step in a systematic way. Phases 2 and 3 account for a more detailed assessment of an early insurgent phase, which could potentially provide operational planners with a more effective solution to influence the decision cycle of a resistance movement's leadership.

FM 3-24 (Modern Mao)

Phasing and Timing: FM 3-24 Counterinsurgency

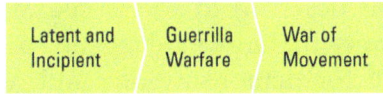

Latent and Incipient	Guerrilla Warfare	War of Movement

Although the three phases of FM 3-24 may accurately represent Mao's model—and may be easier to explain to the general public—such a reduced construct limits the options planners and operators have for force generation. In this respect, simplicity—which is usually something doctrine strives to achieve—may have both positive and negative consequences. If the construct is too complex, it is merely theoretical and cannot be operationalized, but if it is too simple—as the FM 3-24 construct may be—it might not be useful or effective for force generation.

SORO's Five Phases

Phases of Insurgency and Resistance: SORO Phasing Construct

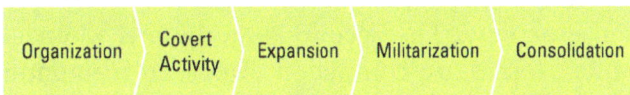

Organization	Covert Activity	Expansion	Militarization	Consolidation

So how, then, can we find the balance between complexity and simplicity? We noted that the SORO publication *Human Factors* provided a strong framework for determinants and variables; can it offer a better phasing construct than Mao, Galula, or ATP 3-05? Perhaps. An effective phasing construct for contemporary resistance exists in the SORO publication, both in the narration, and, later, through analyzing the relationship with the determinants and variables.

From the "Organization Synopsis" section in *Human Factors Considerations of Undergrounds in Insurgencies:*

> To show how the organizational structure of undergrounds changes in protracted revolutions, it is useful to categorize phases in the evolution of conflict. The first phase is the clandestine organization phase in which the underground begins developing such administrative operations as recruiting, training cadres, infiltrating key government organizations and civil groups, establishing escape-and-evasion nets, soliciting funds, establishing safe areas, and developing external support. During this phase, cell size is kept small and the organization is highly compartmentalized.
>
> The second phase is marked by a subversive and psychological offensive in which the underground employs a variety of techniques of subversion and psychological operations designed to add as many members as possible. Covert underground agents in mass organizations call for demonstrations and, with the aid of agitators, turn peaceful demonstrations into riots. Operational terror cells carry out selective threats and assassinations.
>
> In the third or expansion phase the organization is further expanded and mass support and involvement are crystallized. Front organizations and auxiliary cells are created to accommodate and screen new members. During the militarization phase overt guerrilla forces are created. Guerrilla strategy usually follows a three-stage evolution. In the first stage, when guerrillas are considerably outnumbered by security forces, small guerrilla units concentrate on harassment tactics aimed at forcing the government to overextend its defense activity. The second stage begins when government forces are compelled to defend installations and territory with substantially larger forces. The third stage marks the beginning of the full guerrilla offensive of creating and extending "liberated areas." [*Note that this is the FM 3-24/Mao construct.*]

During all of these stages the underground acts as the supply arm of the guerrillas, in addition to carrying out propaganda, terrorist, sabotage, and other subversive activities. Crude factories are set up by the underground and raids are conducted to obtain supplies and weapons. Caches are maintained throughout the country and a transportation system is established. Finances are collected on a national and international basis. Clandestine radio broadcasts, newspapers, and pamphlets carry on the psychological offensive. The underground continues to improve its intelligence and escape-and-evasion nets.

In the fifth phase, the consolidation phase, the underground creates shadow governments. Schools, courts, and other institutions are established to influence men's minds and control their actions, and covert surveillance systems are improved to insure positive control over the populace.[8]

The five-phase paradigm set forth in the SORO publication offers two things. First, it achieves the balance between simplicity and complexity that covers the span of contemporary resistance. Second, and more importantly for practical purposes, this construct enables the operational planning necessary to either enhance or interrupt the growth of a resistance movement at a specific phase.

STAGES OF ORGANIZATIONAL GROWTH: THE MISSING LINK

After concluding that the SORO publication provides the most viable template with regard to both determinants/variables and phase design, the next step in evaluation is to consider whether the model is complete. The organizational diversity of contemporary resistance movements—from individual actors in "leaderless" radical anarchist movements (such as the Occupy Wall Street movement) to strategically organized religious extremist cadres (Islamic State) and criminal-terrorist elements (FARC, Sinaloa Cartel, Boko Haram)—illustrates the complexity of contemporary resistance/insurgent organizations, suggesting the importance between phases of contemporary resistance and the variable stages of organizational growth. No phasing model studied—including the model developed by SORO—adequately covers this variable.

Given this void, it remains unclear which phasing construct is most appropriate for contemporary resistance. To more fully explore which construct best addresses the phases of contemporary resistance, follow-on work must utilize a case-study method to define the stages of organizational growth, analyze how they relate to phases of contemporary resistance as determinants/variables, and compare the advantages and disadvantages of two possible phasing models. Accomplishing this goal requires thorough and detailed study of the stages of organizational growth through case-study modeling that considers existing determinants and variables as they relate to the two proposed phasing constructs.

Proposed Construct A: ATP 3-05 + Organizational Growth

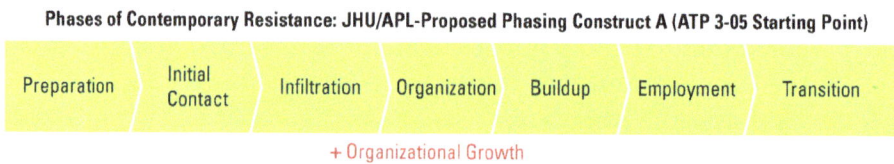

Phases of Contemporary Resistance: JHU/APL-Proposed Phasing Construct A (ATP 3-05 Starting Point)

| Preparation | Initial Contact | Infiltration | Organization | Buildup | Employment | Transition |

+ Organizational Growth

In the current doctrine covering contemporary resistance phases, only ATP 3-05, which appears to be an expansion of Mao's seven phases, presents a model that is operational enough for further evaluation. FM 3-24's three phases and Galula's COIN theory and practice are, respectively, too simplistic and too COIN-centric. Only the ATP 3-05 model offers a template from existing doctrine that could be scaled to either support or deter resistance movements.

Proposed Construct B: SORO + Organizational Growth

Phases of Contemporary Resistance: JHU/APL-Proposed Phasing Construct B (SORO Starting Point)

| Organization | Covert Activity | Expansion | Militarization | Consolidation |

+ Organizational Growth

As the most viable alternative to the constructs in current doctrine, SORO provides the second of the two proposed constructs for phases of contemporary resistance. As with the first, investigation of this phasing construct should include research questions pertaining to the addition of the stages of organizational growth as a determinant and variable, along with an evaluation of the impact of the stage of organizational growth on the model's five phases.

As with the construct put forth in ATP 3-05, it is important to test this five-phase construct against at least two case studies. By adding the stages of organizational growth as a determinant and variable, researchers can access the utility of the construct to provide planners with operational solutions. This assessment will give researchers a sense of the proposed SORO construct's utility versus that of the UW model.

CONCLUSION: PHASES OF RESISTANCE AND FURTHER STUDY

This paper has accomplished three significant steps in developing and validating a construct for the phases of contemporary resistance. First, by reviewing the literature and analyzing existing constructs for the phases of resistance movements, we have deepened our understanding of how current doctrine developed and of other options for phasing. Second, by framing a list of determinants and variables as the primary point of comparison for operationalizing the phases, we have developed a valid testing model for assessing the utility of the phasing construct. Third, we have identified a determinant and variable—stages of organizational growth—that the existing phasing constructs do not consider sufficiently and that require further study.

For additional research, the following next steps merit consideration:

- Confirmation and concurrence with the validity of the existing A and B phase model comparisons
- Identification of contemporary resistance case studies for modeling. In addition to being contemporary, ideally the case studies should account for both existing technology (Internet recruiting, financing through PayPal, psychological operations through Twitter, etc.) and the modern security environment (i.e., after 9/11).
- Careful selection of case studies that are comprehensive enough to cover as many of the prospective phases as possible for both UW and COIN
- Design and validation of the data collection methodology for the case studies
- Definition of a means for measuring the operational utility of phasing construct A versus phasing construct B
- Examination of case study data and recommendation of either construct A or construct B as the most effective phasing construct for understanding contemporary resistance movements

APPENDIX. TABULATED ANALYSIS OF THE VALUES OF VARIABLES RELATED TO INSURGENCY PHASES

Phases of Guerrilla Warfare: Mao Phasing Construct

Determinants and Variables	1. Organization	2. Political Unification	3. Establishing Bases	4. Equipping Forces	5. Recovering National Strength	6. Destroying EN National Strength	7. Regaining Lost Territories
Grievance level	High	High	High	High	High	High	High
Capacity for violence	Low	Low	Moderate	High	High	High	High
Logistics capacity	Low	Low	Moderate	High	High	High	High
Importance of PsyOps	High	High	High	Moderate	Moderate	Moderate	Moderate
Passive resistance	Moderate	High	Moderate	Low	Low	Low	Low
Use of terrorism	Low	Moderate	High	High	High	High	High
Group behavior	Moderately important	Very important	Very important	Very important	Very important	Very important	Very important
Individual behavior	Very important	Very important	Low importance	Low importance	Low importance	Low importance	Low importance
Clandestine behavior	Very important	Very important	Very important	Moderately important	Moderately important	Moderately important	Moderately important
Recruitment	Very important	Very important	Very important	Very important	Moderately important	Moderately important	Moderately important
Education and training	Political ideology	Political ideology	Military logistics	Military logistics	Military operations	Military operations	Military operations
Finance	Very important	Very important	Very important	Very important	Moderately important	Moderately important	Moderately important
Operational orientation	Politically oriented	Military oriented	Military oriented	Military oriented	Politico-military oriented	Politico-military oriented	Terrain oriented
Crowd manipulation	Not important	Very important	Moderately important	Low importance	Very important	Very important	Very important
Leadership importance	Very important	Very important	Very important	Very important	Very important	Very important	Very important

Phases of UW: ATP 3-05 Unconventional Warfare

Determinants and Variables	1. Preparation	2. Initial Contact	3. Infiltration	4. Organization	5. Buildup	6. Employment	7. Transition
Grievance Level	High	High	High	High	High	High	Moderate
Capacity for violence	Low	Low	Moderate	Moderate	High	High	High
Logistics capacity	Low	Low	Low	Moderate	Moderate	High	High
Importance of PsyOps	High	High	High	High	High	Moderate	Moderate
Passive Resistance	High	High	Moderate	Moderate	Low	Low	Low
Use of Terrorism	Low	Low	Low	Low	Low	Moderate	Low
Group Behavior	Low importance	Moderately important	Very important	Very important	Very important	Very important	Very important
Individual Behavior	Very important	Very important	Low importance	Low importance	Low importance	Low importance	Low importance
Clandestine Behavior	Very important	Very important	Very important	Moderately important	Moderately important	Moderately important	Moderately important
Recruitment	Very important	Very important	Very important	Very important	Moderately important	Moderately important	Moderately important
Education and Training	None	None	Military tactics	Military organization	Military logistics	Military operations	Civil governance
Finance	Very important	Very important	Very important	Very important	Moderately important	Moderately important	Moderately important
Operational Orientation	Politically oriented	Politically oriented	Politico-military oriented	Military oriented	Military oriented	Military oriented	Politico-military oriented
Crowd manipulation	Not important	Not important	Not important	Not important	Moderately important	Very important	Very important
Leadership importance	Very important	Very important	Very important	Very important	Very important	Very important	Very important

Phases of Insurgency: Galula Bourgeois-Nationalist "Shortcut" Phasing Construct

Determinants and Variables	1. Organization	2. Blind Terrorism*	3. Selective Terrorism*	4. Establishing Bases	5. Equipping Forces	6. Recovering National Strength	7. Destroying EN National Strength	8. Regaining Lost Territories
Grievance level	High	High	High	High	High	High	High	High
Capacity for violence	Low	High	High	High	High	High	High	High
Logistics capacity	Low	Low	Low	Moderate	High	High	High	High
Importance of PsyOps	High	High	High	High	Moderate	Moderate	Moderate	Moderate
Passive resistance	Moderate	High	High	Moderate	Low	Low	Low	Low
Use of terrorism	Low	High	High	High	High	High	High	High
Group behavior	Moderately important	Not important	Moderately important	Very important	Very important	Very important	Very important	Very important
Individual behavior	Very important	Very important	Moderately important	Low importance	Low importance	Low importance	Low importance	Low importance
Clandestine behavior	Very important	Very important	Very important	Very important	Moderately important	Moderately important	Moderately important	Moderately important
Recruitment	Very important	Very important	Very important	Very important	Very important	Moderately important	Moderately important	Moderately important
Education and training	Political ideology	Terror tactics	Terror tactics	Military logistics	Military logistics	Military operations	Military operations	Military operations
Finance	Very important	Moderately important	Moderately important	Very important	Very important	Moderately important	Moderately important	Moderately important
Operational orientation	Politically oriented	Population oriented	Population oriented	Military oriented	Military oriented	Politico-military oriented	Politico-military oriented	Terrain oriented
Crowd manipulation	Not important	Very important	Very important	Moderately important	Low importance	Very important	Very important	Very important
Leadership importance	Very important	Not important	Not important	Moderately important	Very important	Very important	Very important	Very important

*Replaces "Political Unification" in Mao construct.

Phasing and Timing: FM 3-24 Counterinsurgency

Determinants and Variables	Latent and Incipient	Guerrilla Warfare	War of Movement
Grievance level	High	High	High
Capacity for violence	Moderate	High	High
Logistics capacity	Low	High	High
Importance of PsyOps	High	High	Moderate
Passive resistance	Moderate	Low	Low
Use of terrorism	Moderate	High	High
Group behavior	Moderately important	Very important	Very important
Individual behavior	Very important	Low importance	Low importance
Clandestine behavior	Very important	Moderately important	Moderately important
Recruitment	Very important	Very important	Moderately important
Education and training	Political ideology	Military logistics	Military operations
Finance	Very important	Very important	Moderately important
Operational orientation	Politico-military oriented	Military oriented	Politico-military oriented
Crowd manipulation	Not important	Low importance	Very important
Leadership importance	Very important	Very important	Very important

Phases of Insurgency and Resistance: SORO Phasing Construct

Determinants and Variables	1. Organization	2. Covert Activity	3. Expansion	4. Militarization	5. Consolidation
Grievance level	Moderate	Moderate	High	High	High
Capacity for violence	No	Possible	High	Yes	Yes
Logistics capacity	None	Low	Moderate	High	High
Importance of PsyOps	High	High	High	Moderate	Moderate
Passive resistance	Moderate	High	Moderate	Low	Low
Use of terrorism	None	Selective terror	High	High	Moderate
Group behavior	Not important	Focus on group agitation	Very important	Very important	Very important
Individual behavior	Very important	Very important	Low importance	Low importance	Low importance
Clandestine behavior	Very important	Very important	Very important	Very important	Moderately important
Recruitment	Very important	Very important	Very important	Moderately important	Moderately important
Education and training	Ideology and tactics	Military tactics	Military logistics	Military operations	Governance implementation
Finance	Very important	Very important	Very important	Very important	Very important
Operational orientation	Covert activity oriented	Movement expansion	Military oriented	Military oriented	Politico-military oriented
Crowd manipulation	Not important	Very important	Moderately important	Very important	Very important
Leadership importance	Not important	Moderately important	Very important	Very important	Very important

Phases of Contemporary Resistance: JHU/APL-Proposed Phasing Construct B (SORO Starting Point)

Determinants and Variables	1. Organization Origin	2. Covert Activity	3. Expansion	4. Militarization	5. Consolidation
+ Organizational Growth	?	?	?	?	?
Grievance Level	Moderate	Moderate	High	High	High
Capacity for violence	No	Possible	High	Yes	Yes
Logistics capacity	None	Low	Moderate	High	High
Importance of PsyOps	High	High	High	Moderate	Moderate
Passive Resistance	Moderate	High	Moderate	Low	Low
Use of Terrorism	None	Selective terror	High	High	Moderate
Group Behavior	Not important	Focus on group agitation	Very important	Very important	Very important
Individual Behavior	Very important	Very important	Low importance	Low importance	Low importance
Clandestine Behavior	Very important	Very important	Very important	Very important	Moderately important
Recruitment	Very important	Very important	Very important	Moderately important	Moderately important
Education and Training	Ideology and tactics	Military tactics	Military logistics	Military operations	Governance implementation
Finance	Very important	Very important	Very important	Very important	Very important
Operational Orientation	Covert activity oriented	Movement expansion	Military oriented	Military oriented	Politico-military oriented
Crowd manipulation	Not important	Very important	Moderately important	Very important	Very important
Leadership importance	Not important	Moderately important	Very important	Very important	Very important

Phases of Contemporary Resistance: JHU/APL-Proposed Phasing Construct A (FM 3-24 Starting Point)

Determinants and Variables	Latent and Incipient	Guerrilla Warfare	War of Movement
+ Organizational Growth	?	?	?
Grievance Level	High	High	High
Capacity for violence	Moderate	High	High
Logistics capacity	Low	High	High
Importance of PsyOps	High	High	Moderate
Passive Resistance	Moderate	Low	Low
Use of Terrorism	Moderate	High	High
Group Behavior	Moderately important	Very important	Very important
Individual Behavior	Very important	Low importance	Low importance
Clandestine Behavior	Very important	Moderately important	Moderately important
Recruitment	Very important	Very important	Moderately important
Education and Training	Political ideology	Military logistics	Military operations
Finance	Very important	Very important	Moderately important
Operational Orientation	Politico-military oriented	Military oriented	Politico-military oriented
Crowd manipulation	Not important	Low importance	Very important
Leadership importance	Very important	Very important	Very important

NOTES

1 Mao Zedong, *On Guerrilla Warfare* (Urbana: University of Illinois Press, 2000).

2 Field Manual 3-24 (FM 3-24), *Counterinsurgency* (Washington, DC: Headquarters, Department of the Army, 2006).

3 Fleet Marine Force Reference Publication 12-18, *Mao Tse-tung on Guerrilla Warfare*, transl. Samuel B. Griffith (Washington, DC: Headquarters, Department of the Navy, 1989), 43.

4 Sun Tzu, *The Art of War*.

5 US Army Doctrine and Training Publication (ATP) 3-05, *Unconventional Warfare* (Washington, DC: Headquarters, Department of the Army).

6 David Galula, *Counterinsurgency Warfare: Theory and Practice* (Westport, CT: Praeger Security International, 2006).

7 Andrew R. Molnar, *Human Factors Considerations of Undergrounds in Insurgencies* (Washington, DC: Special Operations Research Office, American University, 1966).

8 Ibid., 2–3.

THE PUBLIC COMPONENT

Summer Newton and Eric Dunford

INTRODUCTION

For most insurgent groups, armed violence makes up only a small portion of its activities. Those individuals that do take up arms within a conflict zone are likewise few, about an estimated 5 percent.[1] The remaining civilians either avoid involvement in the conflict entirely or participate in nonviolent supporting roles. Within insurgent campaigns, violence is only part of the story, but it is the one most often told. Current US Army doctrine, as depicted in previous Assessing Revolutionary and Insurgent Strategies (ARIS) writings, models insurgent organizations based on three primary components: the armed component, the underground component, and the auxiliary component.[2]

Each of these serves a different strategic purpose thought to be critical to the overall success of an armed campaign. The armed component constitutes the visible elements of a resistance movement actively and strategically employing violence through an armed paramilitary force. Armed components typically use a mixture of guerrilla and conventional tactics and primarily concern themselves with the palpable use of violence to achieve their goals. By contrast, the underground component denotes the group's operations that are invisible, or clandestine, such as coercion, strategic manipulation of target populations, spying, and the covert dissemination of propaganda. Finally, the auxiliary component captures the population of noncombatant supporters who do not openly indicate their involvement in or sympathy for the movement but who nonetheless assist the day-to-day operations of the organization. Civilians might act as couriers, provide safe houses, or store weapons for insurgents, all of which are important to the success of the campaign despite being contributions nonviolent in nature.

These three components, which together form the heuristic model the US Army uses to conceptualize insurgent groups, first emerged after World War II. Since that time, however, examinations of insurgent campaigns have begun to highlight a range of additional nonviolent strategies and tactics insurgents use to achieve their goals. The provision of public goods to civilian populations by the Liberation Tigers of Tamil Eelam (LTTE) and Hizbollah and the use of political parties to pursue a political agenda through existing state institutions by the Provisional Irish Republican Army (PIRA) and Hamas speak to possible strategic approaches that fall outside the existing framework of irregular warfare.

Today's insurgencies are more likely to include a public sphere in addition to the armed, underground, and auxiliary components outlined above. The public component refers to the strategic processes that insurgent organizations use to win over, activate, and interact with civilian populations. It is the component of armed insurgencies that is public in nature—that captures an insurgency's capacity to simultaneously engage in violent and nonviolent opposition.[3] Unlike the other components outlined above, the public component offers a strategic approach that is not necessarily violent, coercive, or subversive in nature, but rather one that uses familiar, "state-like" processes and mechanisms to create normalized channels through which governments and civilians alike can interact with an insurgency organization.

It is important to note that these components serve as analytic tools by which to classify and examine the various processes used by modern insurgency groups. The components seek to reduce a complex system into conceptual heuristics in an effort to provide researchers greater analytic leverage when assessing why rebel groups do what they do. That said, how groups use the various components is often more fluid and natural in real life than this definitional framework implies. Thus, the boundary where one component ends and another begins is not always easy to discern on the ground. This paper sets out to make that distinction more palpable by articulating why insurgent groups opt to develop a public component, the strategic benefits of doing so, and the underlying implications and insights that the existence of a public component offers policy makers and researchers alike. In this regard, the paper seeks not to usurp prevailing ARIS doctrine but rather to update it in order to provide a conceptual framework that more appropriately maps onto the empirical situation observed on the ground.

"OLD WARS" VERSUS THE NEW REALITY OF IRREGULAR WARFARE

Since the end of the Cold War, the nature of warfare has changed significantly. Whereas interstate conflict (i.e., a conflict between two or more states) used to be the norm, intrastate conflict (i.e., a conflict occurring within a state's borders, such as a civil war) now dominates. As one researcher notes, of the thirty-seven conflicts occurring worldwide in 2011 alone, only one was an interstate conflict.[4] In addition to this change, how conflicts are resolved has also undergone a dramatic shift. Since the 1990s, more and more conflicts conclude not through military victory but through negotiated settlements, usually resulting in power-sharing agreements that leave both sides militarily intact.[5] As a result, conflicts that end through negotiation tend to be highly unstable. Because of an inability to credibly punish defectors and enforce agreements when both sides are still armed, committing to a negotiated settlement can be problematic, especially as tensions rise. This inability to credibly commit to a negotiated settlement makes conflicts that end in this manner twice as likely to revert back to violence as ones that were resolved through military victory.[6] However, the number of intrastate conflicts that are successfully resolved has greatly increased in the post–Cold War era thanks in large part to the prevalence of negotiated settlements as a viable form of conflict resolution. From 1990 to 2005, a total of 145 intrastate conflicts were successfully resolved, whereas from 1946 to 1990, only a total of 141 conflicts ended.[7]

This shift in the nature of intrastate warfare is widely attributed to some of the larger changes that took place in the international system in the post–Cold War era. After the Cold War's end, the money and resources that came from both the USSR and the United States to fund insurgency operations abroad stopped. The ideological motor that had been the driving force behind many civil wars quickly dwindled, and with the cessation of external support, many insurgency movements found themselves starved both of meaning and of the means to win. In concert with this, the end of a purely bipolar arrangement within the international system created new openings for concerned governments, nongovernmental organizations (NGOs), and international organizations to enter the international scene and promoted conflict resolution and peacekeeping operations within ongoing conflict zones.[8] All these factors had a dramatic impact on the nature of irregular warfare in how it was both fought and resolved.

This distinction between the "old wars" of the Cold War era that were ideologically and geopolitically driven, conventionally fought (i.e., violence was directed toward militaries and other armed actors), and state funded has given away to a new form of warfare. As Mary Kaldor notes, these "new wars" are ethnically, religiously, or regionally motivated and financed either through predatory tactics (i.e., looting); taxing humanitarian groups; or external support through either a "government in exile," diaspora, or some other form of external actor.[9] Kaldor contends that the methods witnessed within these new wars differ from their older counterparts. Traditionally, conventional battles were waged with an eye toward capturing and holding land. Victories were decisive in nature and were directed toward an armed opposition that met on some form of front. However, within the modern context, we see armed groups strategically using civilian populations to gain territory and grow. Modern insurgencies exploit holes in the state's infrastructure and power. Thus, within weaker countries where state power is more difficult to project, especially in rural areas, insurgency groups use a cocktail of violence, public goods, and coercion that is specifically directed at the civilian population in order to capture and hold territory.

The advantages of territorial control for an insurgency organization abound. Groups that are able to hold and control territory possess a greater capacity to coordinate and plan more sophisticated attacks (by having bases) and to expand both militarily (through greater resource extraction) and in size (through recruitment). In contrast, groups that fail to capture or keep territory must operate clandestinely, which often limits the types of tactics they can use (e.g., small explosions, terror tactics, and assassinations) while simultaneously undermining their ability to grow as an organization.[10] In an environment where absolute victory is rare, gaining military and strategic advantages that can offer a rebel organization greater bargaining power during negotiations is often how groups gain and solidify their power in the long term. Capturing territory is one way for a rebel group to display organizational power that directly feeds into the type of concessions the group can demand once peace talks commence.

Holding territory, however, is often complicated by the civilian population already living there. To effectively take over and hold onto a stretch of land, contemporary rebel groups are finding that they must engage in some form of governance with civilian populations. The need to govern is not solely a function of maintaining territory, and at times rebel groups also seek to win over civilian populations through the provision of public goods and services. This process of winning the hearts and minds of a specific population can offer strategic advantages that can also translate into greater leverage at the bargaining table.

In this regard, it is useful to think of rebel organizations and incumbent governments as competing over two types of territory: one that is geographical in nature (such as land, waterways, resources, etc.) and another that is perceptual in nature (such as the approval and support of a population). The conquering of either "space" can directly feed into a group's military strength, increasing its ability to seek greater concessions. For example, groups that enjoy public support are less likely to be politically marginalized, are more able to recruit new members, possess a more robust auxiliary component, and often need to expend fewer resources to govern. However, making the perceptual territory that a group controls known and easy to recognize is notoriously difficult to do, partly because of the fog of war, fears of retaliation or abuse on civilian populations that vocalize such support, and the inherent complications of polling the population to get a reliable and credible measure of support. In this regard, the public component offers a means by which insurgency groups can "publish" civilian support for the organization through the share of votes the group's political party receives. Thus, in many ways, the formation of the public component by insurgency organizations is largely a function of this need to hold geographical *and* perceptual territory.

When assessing how groups use the public component, it is important to stress that not all groups have the same end goals in mind. Some organizations, such as the Lord's Resistance Army in Uganda, are largely predatory in nature. This organization does not seek to gain and hold geographic or perceptual territory; rather, they it is nomadic, largely seeking to extract resources and wealth with little regard for the local population. The group pursues economic predation (i.e., short-term wealth) rather than an eventual spot at the bargaining table (which could be thought of as the pursuit of more long-term gains). In contrast to this, groups such as Hizbollah in Lebanon expend great amounts of energy and resources on developing robust public components with the goal of governance in mind. Thus, it is useful to think of all insurgency groups as falling along a spectrum with respect to how they choose to treat and use civilian populations to achieve their goals. In this regard, the formation and use of a public component varies markedly depending on where groups fall along this spectrum.

PUBLIC COMPONENT AS A "PUBLIC INTERFACE"

The use of the public component by insurgent organizations implies a strategic approach that seeks to engage civil society through socially recognized, nonviolent channels. These are methods of access that incumbent governments and existing civilian populations immediately recognize and understand, such as free social services or political parties. Before coalescing into a fully operational insurgent organization, Hamas's founder, Seikh Ahmed Yassim, concentrated his fundamentalist ideology and efforts into educational programs, charities, social aid, and other humanitarian activities. Acknowledging the limited public support for Hamas's extreme Islamic take on the formation of a Palestinian state, Yassim sought to use the provision of public goods to impoverished Palestinians as a vehicle for gradual indoctrination. In this manner, he was able to pave the way for Hamas's more militant and violent ambitions.[11]

The use of social services and other public institutions to gain support is strategic in nature. Whether it be to activate, win over, or placate civilian populations, the underlying strategic implication is the same: *how* civilians perceive a rebel organization matters. An organization's ability to channel public support can yield positive wartime advantages. Thus, as Yassim's efforts illustrate, to build support, insurgent groups must create channels that allow for daily interaction with civilian populations in order to normalize the insurgent group's presence and to socialize the population to its political goals and ambitions.

With that in mind, the public component should be thought of as a public "interface," or put differently, the means by which illicit, covert organizations interact with their constituencies and incumbent governments (and vice versa). Much like an interface to a computer program, the public component offers a system of access that allows for quotidian interaction with a complex and often misunderstood organization. Given the fractured nature of irregular warfare, the ability to distinguish one organization from another is often more difficult than it may appear to the outside observer. In Syria, the number of groups fighting for power and control has grown rapidly since the conflict's start in 2011. In a bid for relevance among the many groups, some opposition forces have sought to engender support from civilians by offering basic services, such as electricity, water, and stable prices for basic commodities.[12]

How rebel organizations choose to interact with noncombatant populations varies markedly. In Nigeria, Boko Haram's campaign of kidnapping and violence among civilians has created an atmosphere of fear around the organization. However, its use of violence and coercion alone has begun to undermine, rather than reinforce, its efforts to create an Islamic state in Nigeria. As one Nigerian noted after the kidnapping of two hundred girls from a boarding school in northeastern Nigeria, "This thing backfired. You can see the condemnation is both Muslim and Christian; everybody is condemning this. Everybody is turning against the insurgents."[13] The indiscriminate use of violence has slowly begun to isolate Boko Haram among Nigerians. In contrast to this, the Liberation Tigers of Tamil Eelam (LTTE) in Sri Lanka used a mixture of coercion and social services to gain public support among the Tamil population it sought to represent. Using existing NGOs within its northern held territory, the LTTE redirected aid and provision efforts to appear as if they were coming from the rebels. In this manner, LTTE was able to manufacture a level of public dependence for services, which, in extension, translated into support for the organization itself.[14]

The creation of a public interface yields a number of strategic benefits that can help an insurgent organization further its goals outside the traditional scope of the armed component. As noted above, an accessible interface offers insurgents a method by which to distinguish themselves within a chaotic war environment. In addition, by providing goods and services to civilian populations or by representing them through a political party, rebel groups are able to garner greater levels of public support, which can then translate into a greater bargaining ability with an incumbent government in order to solidify any gains that the group has made during the conflict. As the Boko Haram example elucidates, public contempt and scorn can politically isolate rebels, making them more vulnerable to counterinsurgency efforts. A widely hated insurgent group is easier to destroy because the civilian population is less willing to cooperate with it.

Thus, the creation and maintenance of a public interface can be a vital strategic feature in waging an effective insurgency campaign. Doing so allows for everyday interactions with constituent populations, which can help the organization (1) gain public support, (2) gauge that support either through the share of votes the group's political party receives or through daily observation, (3) vet and indoctrinate potential recruits, and (4) gradually push a group's ideology and vision.

In addition, the use of a public component within irregular warfare creates an environment where negotiations and resolutions become possible. Since 9/11, governments have adopted a policy of zero tolerance toward terrorist groups. The act of negotiating with terrorist or insurgent organizations is complicated by the reality that the incumbent government rarely acknowledges the legitimacy of the group it is fighting. This stymies most attempts toward a meaningful resolution to a conflict, because governments often refuse to sit down with or even invite relevant rebel organizations to the negotiating table. However, through the use of a public interface—such as a political party—rebel groups are able to side step this issue. A political party can offer a legitimate face to an illicit organization, making government interaction possible. More importantly, it creates a means by which both sides can negotiate without either side appearing weak.

Conceptualizing the public component as a public interface offers a way to think about insurgent activities that fall outside the traditional realm of irregular warfare. In doing so, it is possible to articulate a gray area within modern conflicts where rebels become tacit governments or nascent politicians. There are two specific types of public interfaces that insurgent organizations can create: a "goods-based" interface and an "institution-based" interface. A goods-based interface is created when rebel groups selectively provide goods and services to a targeted population with the intention of gaining public support. As captured in previous ARIS analysis, this type of interface falls within the realm of "shadow governance," where rebel organizations actively use the carrot versus the stick in shoring up public support.[15] Existing simultaneously with or independently from this, an institution-based interface manifests when rebel groups create or use existing political parties to engage with the political framework of the state. As will be expounded on below, an institution-based framework can manifest in a number of different ways.

The following two sections will further explore these two types of public interfaces, articulating the unique role each interface can have and the method and manner in which they can vary. It is essential to note that the types of interfaces a group can use are not restricted to one or the other. Some groups, such as the LTTE, only offer public goods, while other groups, like Euskadi Ta Askatasuna (ETA) in Spain, only have a political party. There are, however, groups that opt to have both, such as Hizbollah in Lebanon. The significant variation in how and why insurgent groups choose to use or not use the public component to achieve their strategic goals offers important insights into what the formation of a public component can mean for a group's eventual success.

GOODS-BASED INTERFACE

The goods-based interface of the public component manifests through a rebel group's willingness and propensity to selectively provide goods and services to a targeted population. In an effort to win over relevant populations, insurgent groups offer benefits and other public goods that are designed to normalize civilian interaction with the organization. Communities come to depend on these services, granting civilians the opportunity for quotidian interaction with the rebel organization. As previous research highlights, the power a social service provider has over the population it provides for is contingent on the recipient's ability to seek those services elsewhere.[16] Within conflict zones, as the security situation deteriorates and the state loses its capacity to provide for citizens in rebel-held territories, insurgent groups often become the sole provider of public goods and services within the areas under their control. Thus, the creation of a goods-based interface is often as much a product of necessity as it is of strategy.

Public goods are any goods and/or services that are nonexcludable (i.e., if it is given to one, it is given to all) and nonrivalrous (i.e., if one person uses it, the total quantity is not diminished). Air, national defense, and public streetlights at night are all forms of public goods. When provided by the state, these forms of goods often manifest as social services, such as health care, education, and security services. They are expected features of developed states, and the provision of such goods feeds directly into the legitimacy that states seek to manufacture and maintain.

During times of conflict, the state's ability to provide public goods in war-torn areas is often severely hindered by the ensuing violence.

More often than not, as rebel groups capture territory, the state's juris-diction becomes fractured. Civilian populations within conflict zones are characterized as existing in one of three zones of control: areas under the control of the state, areas of divided or contested control, and areas under insurgent control.[17] When control of a territory shifts from the state to an insurgent group, the state's ability to project its power ceases and the care of the civilian populations falls under the rebel group's jurisdiction. However, how rebel organizations choose to manage this responsibility varies markedly.

Upon gaining territory, some rebellious groups have no intention of engaging the existing population. Often groups seek to depopulate local communities living in controlled territories in order to circumvent the costs that come with governing. After capturing territory in north-ern Rwanda, the Rwandan Patriotic Front (RPF) strongly distrusted the Hutu communities in the areas it controlled. Rather than invest valuable resources in policing and coercing them into submission, the RPF forced the civilian communities out.[18]

Other times rebel organizations opt instead to provide a compre-hensive range of services to the civilian populations under their control. Shortly after its formation, Hizbollah began pouring large amounts of resources into providing public goods to the Shiite community in southern Lebanon. The group offered a whole range of services from social support and childcare networks to comprehensive education and health care systems.[a]

Why some groups choose to provide for populations while others do not depends largely on the kind of campaign being waged. For secessionist and ethnocentric groups, for which the formation of an independent state is the primary goal, winning over relevant portions of the population is essential to their overall success. Brute force and terror can keep populations in line for a time; however, oppression is poorly equipped to win over the hearts and minds of a population. As one researcher notes, "Fear alone does not suffice to sustain rule in the long term."[20] Furthermore, violence and coercion as a group's sole form of governance can have lasting implications as the organization chooses to expand.[21] In addition to this, discontented citizens living under rebel rule can also pose a number of other challenges for rebel leaders, often choosing not to comply with rebel directives or even seek-ing to collaborate with the state against the organization itself.[22]

Thus, groups selecting to control territory face the very real real-ity of having to provide for the populations under their jurisdiction.

---

[a] Although the exact date when Hizbollah began providing public services to poor Shiite communities is unclear, providing social services to the poor has largely been one of the group's key missions since its inception in 1982.[19]

Within this context, support from relevant portions of the population may become strategically necessary. Groups neglecting to take this fact into account may find themselves expending valuable resources to maintain internal order. In the case of the Shining Path in Peru, the organization created such an atmosphere of fear among the civilian population living in the countryside that the government was eventually able to arm peasant militias to fight against the group, which played a key role in the Shining Path's eventual demise.[23]

Shawn Flanigan notes how public support for a rebel organization exists along a continuum, where on one end of the spectrum, civilians actively resist the rebels (as was the case with the Shining Path), and on the other end, civilians choose to actively participate with the insurgent organization as "community members."[24,b] Naturally, a given population of citizens will fall differently along Flanigan's continuum, with some supporting the incumbent government and others the rebels. The key here, however, is the tipping point—below a certain threshold of popular support, governing a civilian population can be more work than it is worth. Thus, groups choosing to take and keep territory have an incentive to provide for the population under their control. How insurgent organizations go about this, though, differs greatly from group to group.

From its beginning, Hizbollah sought to develop its public infrastructure alongside its military one in an effort to "strike roots" within the poor Shiite community in southern Lebanon.[26] Drawing large amounts of financial and logistic support from Iran, Hizbollah's mission was crafted with an eye toward winning over the Shiite population residing in Lebanon. Providing public goods to the impoverished community was part of the essential mission of the organization, which termed the endeavor an act of jihad. At the time, Hizbollah was competing with another Shiite organization, Amal, for the Shiite community's support. As a means of differentiating itself as an organization, Hizbollah began pouring considerable time and resources into social service provisions, which quickly helped increase the group's popularity while simultaneously highlighting its organizational competence within a notoriously corrupt and ineffectual Lebanese state.[27]

Hizbollah's social service system was both well structured and expansive, providing a generous range of social, educational, and health services. The organization created institutions designed to maintain

---

b Civilians who support an insurgent organization to the degree that they are able to assist the group as community members is in large part the essence of the auxiliary component. Civilians aiding the organization either administratively, logistically, or through some other capacity, while simultaneously retaining their role as noncombatants, largely make up this aspect of an insurgency's operation.[25]

southern Lebanon's crumbling infrastructure—building roads, paying reconstruction compensation after Israeli bombings, and fixing and maintaining Beirut's water mains.[28] In addition to this, the organization's provision of health services and "social insurance" reinforced its role as a caretaker of the Shiite people.[29] Hizbollah's creation of an extensive goods-based interface directly fed into the organization's success and helped stabilize it within a rapidly changing political environment after Israel withdrew from southern Lebanon in 2000.

The amount of energy and resources that Hizbollah invested into developing such a robust goods-based interface was partly due to the group's need to differentiate itself as an organization and partly due to the fact that such investment was a central aspect of the organization's mission; however, external backing from Iran played a large part in Hizbollah's ability to fund its service initiatives, potentially making it the exception rather than the rule. Because of funding constraints, not all groups opt to build their goods-based interface from the ground up, and many instead choose to co-opt existing institutions for their own purposes.

Upon acquiring the largely Tamil territory in the northern part of Sri Lanka, the LTTE was faced with a population accustomed to public services that had previously been provided by the incumbent state. Thus, there was an expectation that the LTTE would follow suit; however, without an external backer, building a service system from the ground up would have come at significant cost. Instead, the Tamil Tigers sought to redirect existing institutions and nongovernmental agencies to meet the LTTE's needs.[30]

Deeply suspicious of the Sinhalese government's use of rehabilitation and development programs as a means of undermining the LTTE's authority in the north and east, the organization actively redirected government-provided services to appear as if they were coming from the organization.[31] In addition to this, the LTTE levied taxes against NGOs operating within its territory while simultaneously steering where they were able to operate. As one Sri Lankan noted, "The LTTE does not provide services; instead they allow aid and relief NGOs to operate in their areas, while controlling where and how they operate."[32]

By requiring NGOs to conduct their work through the Tamil Relief Organization (TRO), the LTTE could direct the flow of humanitarian activities in a manner that made it appear as if the services were coming from the LTTE. Their efforts in this regard helped boost the LTTE's legitimacy in the eyes of the Tamil community.[33] Through the use of the TRO and the services it provided, the LTTE was able to ensure public dependency on the organization, especially in the wake of the 2004 tsunami.[34] The strategic manipulation of existing services aided the LTTE in creating a goods-based interface that boosted its standing among its civilian population while providing the organization with a much needed form of revenue. In addition, use of a goods-based interface helped engender a sense of loyalty and community among the Tamil youth, creating a sense of statehood, which reduced the LTTE's need to use coercion to maintain order.[35]

For the LTTE, the strategic benefits of developing a goods-based interface were clear. The organization's dream of statehood prompted it to focus on the Tamil population under its jurisdiction. By steering existing institutions and organizations, the LTTE was able to manufacture a public component that brought all the strategic benefits at limited cost. Furthermore, its efforts in doing so limited the need to use coercion and violence to maintain order. However, not all insurgent organizations that strive to create a goods-based interface are able to do so successfully.

From its beginning, the Rally for Congolese Democracy (RCD) in eastern Congo was either uninterested or unable to do what LTTE and Hizbollah did. The organization's brutal use of violence to maintain order among its civilian population and the internal fissures within the organization itself prevented the RCD from successfully manufacturing a goods-based interface. Much like Hizbollah, the RCD had an external backer; however, this forced the organization to myopically focus on taking the capital of the Democratic Republic of the Congo, Kinshasa. This focus caused the RCD to neglect the populations living within the territory under its control and to overlook the diverse ethnic and political tensions that threatened to rip the insurgent group apart. Relying on largely defunct institutions and aid organizations, the insurgent group left the civilian population to largely fend for its own social services while continuing to maintain its authority and rule through coercion and violence.[36] Thus, the RCD's wish to rule was not coupled with the wherewithal to adequately provide the services for the population under its control, which limited the organization's capacity to push on and achieve its organizational goals.

As these examples highlight, a goods-based interface can come in many forms and how organizations seek to develop it varies greatly. However, when implemented effectively, a goods-based interface can help insurgent organizations construct systems of interaction with civilian populations, which can eventually translate into increased levels of support and legitimacy for the group.

INSTITUTION-BASED INTERFACE

An institution-based interface of the public component manifests when rebel groups create or use existing political parties to engage with the political framework of the state. The goal of an institution-based interface is to create a legitimate face for the insurgent organization that allows it to move and operate within the confines of civil society. Much like a goods-based interface, an institution-based interface seeks to normalize civilian and government interaction with the rebel organization; however, it does so in a much different capacity. With its focus on manufacturing legitimacy among both its constituents and opponents alike, it is a form of interface that exists through the political institutions of the state, offering governments and citizens a means to legally and credibly interact with an illicit organization.

The development of an institution-based interface becomes an important feature of irregular warfare as conflicts begin to draw to a close. In an attempt to stay politically relevant, armed organizations must discover new avenues of operation that extend beyond the mere use of violence. Groups that can anticipate the need for a legitimate political presence reduce their likelihood of being marginalized as the nature of the conflict changes. For example, shortly after the signing of the Oslo Accords and the creation of the Palestinian National Authority (PNA) in 1993, Hamas quickly realized the importance of creating a legal Islamic party that would allow it to operate without being persecuted by security forces within the PNA.[37]

However, the use of political parties by insurgent groups is not mono-directional. Sometimes political parties create and use violent groups to achieve their own goals. The idea of political parties using violence to achieve specific political goals within modern democracies is not a foreign concept. In India, Hindu party leaders regularly set off anti-Muslim protests and riots to shore up support. By making ethnic issues more salient around election times, Hindu parties are able to rally voters through the targeted use of mass violence and unrest.[38] Examples of the strategic use of political violence abound. In Indonesia, Bangladesh, and Sri Lanka, party-inspired ethnic and religious animosities are frequently used to help rally voters or remove opposition candidates.[39]

The use of an institution-based interface is thus more nuanced than its goods-based counterpart. A goods-based interface emerges only after a group's armed component has acquired territory; in contrast, an institution-based interface can actually predate a group's armed component, as was the case with Umkhonto we Sizwe and the African National Congress (ANC) in South Africa. Thus, when conceptualizing and assessing the role an institution-based interface plays within an armed insurgency's strategic thought process, it is important to understand who is calling the shots (the political party or the rebel group) and how that dynamic can change over time.

In political science and economic research, a principal–agent framework is often used to articulate this kind of dynamic between two actors. A principal–agent relationship occurs when one actor (the agent) performs tasks and makes decisions on the behalf of another actor (the principal). Operating in a world where one actor may have more information than the other and where each side has different preferences and methods, the principal cannot always ensure that the agent will operate in its best interest. Thus, the principal may reel the agent in, punish the agent, or break relations with the agent altogether if it fails to perform. Thinking of insurgency groups and political parties within a principal–agent framework offers a way of conceptualizing the dynamic relationship between a rebel organization and the legal political entity that represents it.

When examining the variation in the types of institution-based interfaces that may exist, it is important to assess how the principal–agent dynamic plays out between the two actors. A rebel group that manufactures a political party to gain access to a political arena differs greatly from a political party that creates a military wing to further its political agenda. With this in mind, three types of relationships emerge between rebel groups and political parties: (1) political parties exist as "fronts" when the political party is subordinate to the insurgent group; (2) political parties exist as "principals" when the rebel group is subordinate to the political party; and (3) political parties exist "independent" of the rebel group when neither the party nor the group wields any level of control over what the other does. Understanding how these three possible arrangements work is key to assessing the strategic value of an institution-based interface.

Political Parties as Fronts

Within the first type of arrangement, a political party exists as a front that a rebel group uses to gain access to a state's political arena. Within this setup, the political party has no agency or ability in and of itself to restrain the rebel organization from acting. Rather, the insurgents control (at least partially) the actions and agenda of the party in order to achieve specific goals. Under this arrangement, a political party might act as the legitimate face of the organization at the bargaining table; however, representatives of the political party often report back to the rebel organization to ensure that the group agrees.

Thus, within this arrangement, the political party can be thought of as the agent and the rebel group the principal.

In the beginning, the Irish Republic Army (IRA) and Sinn Féin, a civilian political party, were largely separate and independent organizations seeking the unification of Ireland. However, in 1949, the two organizations' paths became inextricably linked as IRA officials began "infiltrating" Sinn Féin and taking up key positions within the political party.[40] From this point on, dual membership in both organizations was not uncommon. As one senior Irish official noted, membership in both the IRA and Sinn Féin was the "norm rather than the exception, particularly at the highest levels" of the party.[41] As the conflict endured in Northern Ireland, the IRA's Army Council, the main level of leadership within the IRA, was largely assumed to have control of Sinn Féin's political agenda.[42]

In 1969, the nature of the conflict in Northern Ireland changed. As the cleavage between Protestants and Catholics became more salient, British troops were sent in to restore order between the two groups. However, the British military presence on the island altered how the conflict was viewed: many began to see British occupation as a barrier to the unification of the island.[43] Around that time, a new, more violent breed of membership emerged within the IRA that eventually formed into the Provisional IRA (PIRA). This new faction of Irish Republicans relied largely on violence to achieve its goal of a united Ireland. However, some leaders within the PIRA began to argue that dependence on violence alone would only succeed in isolating the rebel organization.[44] The PIRA sought to transform Sinn Féin into a fully fledged political party and attempted to achieve its goals through both conventional political channels and military offensives. From this point, Sinn Féin began participating in local elections. In this regard, PIRA relied on the ArmaLite and the ballot box, or the so-called "ballot bomb," to achieve the organization's goal of unification.[45]

The use of the political framework of Northern Ireland played a key role in helping the PIRA gauge its level of public support. Through the ballot box, the PIRA was simultaneously able to effect local legislation and understand where it stood among its constituents. For violent and covert organizations, it is often difficult to gauge public support. What that support means differs from group to group; however, as was noted above, public support can directly feed into an organization's ability to achieve its goals. Thus, the ability to poll a group's current standing through the use of timely elections can be as important as the seats it wins. For the PIRA, Sinn Féin offered a means by which to navigate the shifting political landscape of Ireland at the time—a fact that would prove vital as the conflict drew to a stalemate in the late 1980s and peace negotiations commenced.

When negotiations began in 1994, many international actors promoting peace in Northern Ireland thought it best to keep the relationship between Sinn Féin and the PIRA as murky as possible.[46] The British government concluded that it was pointless to exclude political parties with ties to paramilitary organizations from the peace talks, knowing that doing so would only succeed in derailing the process.[47] Thus, Sinn Féin was invited to the negotiating table under the tacit understanding that the PIRA would be participating in the negotiations. During the negotiations, leading members of Sinn Féin reportedly ferried messages back to the PIRA's leadership.[48] After declaring a cease-fire in 1997, Sinn Féin was kicked out of the negotiations when it surfaced that the PIRA played a role in a recent spate of terrorist attacks at the time. As this point illustrates, Sinn Féin operated largely as a front for the PIRA during the negotiating process. Seeking an end to the conflict, the British government and other actors opted to implicitly negotiate with the PIRA. Sinn Féin merely made this process possible.

The strategic use of Sinn Féin as an institution-based interface was central to the PIRA's ability to stay relevant within a shifting political landscape. As conflicts draw to a close, possessing some form of legitimate interface is integral to keeping an organization from being marginalized or isolated. When violent campaigns draw to a standstill, insurgent groups must often enter into negotiations in order to solidify any gains made during the conflict. As the PIRA example demonstrates, the capacity in which this process of negotiation can take place is only made possible through the existence of an institution-based interface.

Like the PIRA, in 1993 with the signing of the Oslo Accords, Hamas faced the very real threat of being politically isolated within the new governing arrangement that brought about the creation of the PNA. In an attempt to reconcile its hard-line Islamic ideology and to gain influence among the Palestinian population residing in the West Bank, Hamas decided to enter the political arena, forming the Islamic National Salvation Party in 1995. By forming a political party, Hamas sought to pursue its vision for an Islamic state by engaging existing channels in order to influence the domestic political arena.[49] The party was a marginal player within Palestinian politics until the turn of the century. However, through the creation of a goods-based interface, Hamas was able to strengthen its ties with the Palestinian community in the West Bank. Its efforts in this regard translated into a stunning victory in 2006, winning Hamas 74 of 132 seats within the Palestinian legislature.[50] In 2007, Hamas's forces took over the Gaza Strip, effectively dismantling the Palestinian Authority's military forces and purging the political landscape of challengers.

Hamas's use of the Islamic National Salvation Party to gain traction within a tumultuous political and security landscape was purely strategic. Through the formation of an institution-based interface, Hamas was able to access the political arena in an effort to win over much needed public support and recognition. Fearing isolation because of its strong ideological stance, Hamas's strategic use of an institution- and goods-based interface allowed it to advance politically, putting the organization in a position to undermine challengers, who years before had sought to isolate the organization. The Islamic National Salvation Party was a front for Hamas's military initiative, intimately tied to the insurgent organization itself. Through its political party, Hamas was able to advance militarily in a manner that would have proven impossible through armed violence alone.

Political Parties as "Principals"

In the second type of arrangement, rebel groups are subordinate to the political party. The political party within a principal arrangement is the primary decision maker, whereas the armed group serves as its agent. Traditionally, this type of dynamic manifests when political parties opt to have a violent wing in response to institutional restrictions, an inability to gain office via election, the state's banning of the party altogether, or government repression. Under these conditions, a political party views the existing political framework of the state as inaccessible and thus attempts to change the status quo through alternative means. The formation and use of an insurgent organization can be a controlled effort or an attempt to channel a radical faction within the party itself. Either way, the principal arrangement holds as long as the party is able to wield influence over how and where violence is used.

Initially created in 1912, the ANC sought to bring equal political rights to South Africa. However, as a series of laws were put in place to segregate black communities within the country, the ANC turned to nonviolent resistance against the incumbent government. In 1948, an apartheid government was put in place, denying nonwhites access to the institutional framework of the state. For example, all nonwhites were removed from the electoral roll, preventing them from participating in general elections. When the ANC joined the Defiance Campaign of Unjust Laws, a transnational movement bent on curbing unjust legislation, the apartheid regime reacted harshly by arresting all the ANC's leadership and initiating a number of "Treason Trials."[51] When the police opened fire during a nonviolent demonstration in 1960, protests

41

and strikes against the government increased, prompting the regime to initiate a state of emergency, making all nonwhite organizations illegal.

The ANC's inability to access the institutional framework of the state in order to change the status quo forced it into active protest. However, its campaign of nonviolent demonstration proved ineffective at deterring the state from its violent and repressive course. The ANC leadership quickly realized that to achieve its goals, it needed an ability to fight violence with violence. The political party created a military branch called the Umkhonto we Sizwe ("Spear of the Nation," or MK) with the intent of engaging in acts of sabotage against the apartheid state. Through the strategic and controlled use of violence, the ANC sought to use the rebel group as a tool for change. Although membership between the two organizations was often blurred, the ANC took great care to keep the operations and activities of the ANC and Umkhonto separate. As one researcher notes, "The ANC remained the political apparatus responsible for carrying out all political activities, while Umkhonto focused on armed resistance. Maintaining separate organizations, with distinct policies, meant that the ANC never officially adopted a policy of violent resistance."[52] Given the repressive apparatus of the state and the nonviolent nature of the movement itself, it was integral that the two groups remained separate, even if the ANC was still calling the shots.

The ANC's formation of an armed component offers a reverse logic to how violent nonstate groups form within the context of irregular warfare. As the example illustrates, the public component may exist long before the other components of an insurgent group emerge. This highlights the conceptually tricky gray area that the principal arrangement of an institution-based interface occupies. How the various components are used matters only with regard to who is controlling the violence. In this regard, a principal arrangement is unique. However, not all principal arrangements emerge in the same manner.

Some political parties form as fronts but then slowly begin to gain in power and importance as an armed conflict subsides and civil society is reintroduced. In this context, an ad hoc principal arrangement can emerge where a front arrangement previously existed. After the death of Ayatollah Khomeini in 1989, Hizbollah was offered greater freedom to pursue its own goals separate from the edicts coming from Tehran. The organization decided to participate in the Lebanese national parliamentary elections in 1992. The choice to create a political party was strategically motivated. Hizbollah's leadership believed that engaging Lebanon's political system would offer it a voice in the drafting of any future political arrangement with Israel—which it hoped to derail.[53] More importantly, Hizbollah understood that only through political channels could it ensure the organization's survival even if it was called to disarm. Lastly, as a unified Lebanese state began to emerge after twenty years of civil war, Hizbollah sought access to lucrative governmental positions and resources that it could only activate via legitimate channels.[54] Thus, the creation of a political party was Hizbollah's way of responding to a rapidly shifting political environment in Lebanon at the time.

In the 1992 election, Hizbollah ran with Amal, another Shiite group that was the more popular of the two organizations at the time. The coalition won 8 of the 128 seats in the parliament that year.[55] Bolstered by its political success, Hizbollah began expanding its goods-based interface to increase its political base in an effort to further its political pursuits. The organization's ability to use a goods-based interface through the provision of social services in tandem with its creation of an institution-based interface played a large part in its ability to rally grassroots supporters. In the subsequent election, Hizbollah was victorious again.

Israel's withdrawal from southern Lebanon in 2000 increased the strategic importance of Hizbollah's institution-based interface. With the end of Israeli occupation, the organization's armed component came second to the public component it had put in place. Hizbollah won twelve seats in the 2000 election, and then fourteen in 2004, as its officials began gaining key governmental positions within the executive cabinet. As Hizbollah the political party's influence increased, the need for Hizbollah the armed group began to diminish, effectively changing the nature of the principal–agent dynamic from a front to a principal arrangement. Nonetheless, Hizbollah maintained its armed component, using it at times to influence Lebanese politics. In May 2008, Hizbollah staged a six-day takeover of the capital in Beirut in order to gain further political concessions—which it succeeded in doing, effectively winning veto power within the executive cabinet.[56]

Hizbollah the political party started as an organization bent on influencing the postconflict Lebanese political arena; however, its political success at the ballot box only underscored its strategic importance to the organization's leadership. As its need for violence diminished, Hizbollah's institution-based interface became the dominant actor in relation to its armed component. Thus, after 2000, the political party was largely calling the shots, using its armed wing only when the need arose. The transformation of Hizbollah's institution-based interface from a front to a principal arrangement highlights the dynamic ways in which a political party and insurgent group can interact and how that interaction can change over time.

Political Parties Independent of Insurgent Groups

The third form of arrangement seeks to elucidate instances when an institution-based interface is either no longer used by an insurgent group or never existed to begin with. A political party is "independent" of a rebel group when neither the party nor the group wields any level of control over what the other does. In effect, the principal–agent dynamic is no longer at play and the two organizations can be thought of as separate entities. These are parties that are thought to be affiliated with an insurgent organization either by standing for the same cause or representing the same ethnic group but that hold no real influence over the rebel groups' agenda, actions, or goals and vice versa.

When the Kurdistan Worker's Party (commonly known as the PKK) first emerged in the late 1970s in Turkey, the Turkish state responded with more than just a counterinsurgency campaign. The government implemented a range of laws that restricted basic rights in crisis zones where the conflict was most fierce while simultaneously condemning any political party or group that sought to represent the Kurdish cause.[57] By equating national unity with ethnic homogeneity, the Turkish state had a long history of coercive assimilation of non-Turkish groups; however, the PKK's bloody campaign for an independent state in the southeast part of the country exacerbated this tendency.[58] The People's Democracy Party (HADEP), a Turkish political party that stood for greater political rights for Turkey's Kurdish minority, was outlawed and its leadership arrested because of alleged ties to the PKK.[59] The Turkish state saw connections that were largely nonexistent in an effort to purge the political arena of groups that threatened the state's national unity.

Just as an institution-based interface can be used to help further an insurgency's goals, it can likewise be used by states to rid themselves of political competitors. Independent arrangements often emerge within this kind of setting. By purposefully finding patterns where none exist, governments often manufacture narratives of insurgent groups and political parties working together in order to isolate or remove them from the political arena. Thus, when examining the arrangement of an alleged institution-based interface, it is key to take into account historical circumstances that might motivate such a narrative.

An incumbent government seeking to push out competitors is not the only way an independent arrangement emerges. Some insurgent groups seek to manufacture ties with political parties in order to remain politically relevant. In the early 1950s, a number of violent sectarian conflicts erupted between Conservatives and Liberals in Colombia. In response, Gustavo Rojas Pinilla[c] organized a political movement that pursued a populist agenda that sought to underplay ideological difference and emphasize state unity. He formed the National Popular Alliance (ANAPO), which became a key political contender in Colombia. However, during the 1970s election, Rojas's bid for the presidency failed. Because Rojas lost by a margin of 1.6 percent, many loyal to Rojas believed the election results had been manipulated.[60] Unable to accept defeat, Rojas's loyalists formed the Movimiento 19 de Abril (M-19), which took its name from the April 19 election date on which Rojas lost. The rebel organization termed as the military wing of the ANAPO; however, the ANAPO detached itself from the organization. When ANAPO's leadership, which was spearheaded by Rojas's daughter Maria, took steps away from the M-19, the rebel organization continued to claim that it was the party's military arm.

In reality, the insurgent organization, partly composed of leftists and disgruntled FARC members, became largely irrelevant to the ANAPO and thus did not serve the same kind of strategic usefulness that Umkhonto did for the ANC in South Africa. Sustaining ties with the M-19 yielded no real political benefit, especially within Colombia's rapidly changing political environment. However, with no real ideological position, the M-19 needed the ANAPO to remain relevant as an organization. The relationship between the M-19 and the ANAPO highlights how certain arrangements can deteriorate.

In effect, an independent arrangement is a noncategory with regard to how an institution-based interface is used to further a rebel organization's goals; however, it is important to distinguish it from the other types of arrangements. Being able to distinguish a front or principal arrangement from an independent one changes how one assesses the strategic utility of the public component being used. Only through understanding the principal–agent dynamic can the strategic use of the public component become clear.

THE STRATEGIC VALUE OF THE PUBLIC COMPONENT

As noted above, an insurgent group's creation and implementation of a public component stems largely from the strategic benefits of doing so. The underlying assumption captured in this analysis is that the public component becomes strategically relevant as the likelihood

---

[c] Gustavo Rojas Pinilla was a military general who ruled Colombia after mounting a successful 1953 coup d'état there. He sought to restore order to the country, which was beset with a bloody partisan civil war, known as La Violencia (The Violence), in the countryside. He was deposed in 1957 after public protests against his rule.

of absolute victory begins to diminish. As groups realize that their political goals (be they secession, regime change, the reconfiguration of the rule of law, etc.) are no longer achievable in absolute terms, the need for an alternative interface with the government and public becomes strategically necessary to solidify any gains the group has made and to keep the organization from becoming politically isolated. Hizbollah, Hamas, and the PIRA all formed institution-based interfaces in response to changing political circumstances. Specifically, the PIRA noted during the height of The Troubles that violence could yield a form of legitimacy but it did not offer legality, and that to reach its goal of a unified Ireland, the group needed legality.[61] Thus, the creation of a public component can be seen as a process of strategic adaptation, where rebel groups begin to use a number of different strategies simultaneously to generate effective campaigns that are more adept at achieving the group's goals.

When used in concert with the armed, auxiliary, and underground components, the strategic benefits of the public component abound. The use of a public interface (through a political party, the provision of public goods, or both) serves three unique purposes for rebel groups: it offers a legitimate face that both the government and the public can credibly interact with, which becomes essential during negotiations; it shores up public support through the provision of goods and services; and it offers a means to publish public support for the organization.

By creating processes by which civilians can interact with the organization and services that make that interaction meaningful, rebel groups strategically attempt to win over select portions of the population to distinguish themselves, garner greater levels of cooperation, and generate a larger pool of recruits. Public support can manifest in a number of ways. It can emerge politically through the share of votes the group receives through a political party or it can emerge pragmatically when the group needs either resources or other forms of assistance. In this regard, the type of environment insurgents are fighting in matters. In resource-rich countries, resources are more readily available, but in resource-poor countries, rebels must appeal to civilian populations in order to extract valuable resources and recruits.[62] To wage a successful campaign, groups require the guns, people, and provisions necessary to keep the fight going. However, when resources are more difficult to come by, groups must find ways to garner resources from civilian populations. In this regard, the public component is the strategic use of the carrot versus the stick in drawing support and resources from the civilian population. Furthermore, an insurgent group is more able to disappear into a civilian community that supports it, offering it a means by which to avoid state-led counterinsurgency efforts.

In addition to this, the creation of a public interface offers insurgent organizations the means by which to find, vet, and indoctrinate new recruits. Researchers have pointed to Hamas, Hizbollah, and the LTTE's use of charities as a mechanism to socialize and recruit new members, often using services aimed at rewarding fighters and their families.[63] Through a goods-based interface, rebel groups are able to interact with civilian populations on a more regular basis. Through health care and social services, rebel leaders are able to isolate potential recruits and systematically indoctrinate them.[64] Thus, the creation of a public interface eases these forms of interaction by normalizing them.

Through a political party, insurgent groups are able to publish the level of public support they receive in an attempt to strengthen their position at the bargaining table. Through the use of an institution-based interface, rebels have a means of polling civilians on a semiregular basis. As the level of public support a group receives grows, the group is able to demand more from the incumbent government and other competing parties. For example, after winning multiple elections, Hizbollah was able to use a mixture of politics and coercion to seek greater concessions from the Lebanese government. In this regard, groups that are able to project levels of support through a legitimate medium are more able to achieve their military objectives by better articulating their bargaining space.

Finally, through the creation of a public interface, insurgent groups increase their likelihood of being able to reach a negotiated settlement. As noted above, governments are rarely willing to speak directly to rebel organizations themselves; thus, political parties act as an interface through which negotiations can occur without either side appearing weak. By creating credible channels of interaction, rebel groups are more able to solidify political concessions and resources. Thus, the public component offers a means by which to credibly exit a conflict along negotiated terms and conditions rather than through capitulation, or worse, absolute defeat. When the prospect of absolute victory is no longer a reality, rebel groups face the dismal prospect of fighting a war of attrition or losing altogether. Within this context, the ability to facilitate and reach a negotiated settlement offers groups a means of ending the conflict in a position of power.

Furthermore, through a political party, rebel groups are able to bargain indirectly through the institutional framework of the state. By participating in the political arena, groups are able to influence the policy outcomes from which they would otherwise be barred. Thus, the use of an institution-based interface can potentially yield both a direct and indirect means of resolving a conflict.

CONCLUSION: THE PUBLIC COMPONENT WITHIN EXISTING DOCTRINE

In an attempt to provide a heuristic model for irregular warfare, current ARIS doctrine seeks to articulate the various mechanisms that appear—at least empirically—to be most prevalent within modern intrastate conflicts. The theoretical partition of modern insurgencies into armed, underground, and auxiliary components offers analysts a method and means by which to assess, compare, and determine the likelihood of success of existing violent nonstate groups. However, the model's empirical purchase stems primarily from its ability to map onto the situation being observed on the ground.

By introducing a new conceptual component to the current doctrine of irregular warfare, this paper seeks not to obfuscate prevailing attempts at systematically analyzing armed insurgencies. Rather, it attempts to expand on previous work by offering a more robust model that more appropriately speaks to the situation on the ground. In this regard, it is key to articulate how the public component fits into existing doctrine and why the incorporation of the public component into current military understanding of armed groups offers a more complete heuristic model when examining modern conflicts.

As outlined above, the universe of rebel activities that fall within the realm of the public component differs greatly from the established armed, auxiliary, and underground components. By definition, the public component constitutes a strategic approach that seeks to engage civil society through socially recognized, nonviolent channels; thus, as a theoretical category, the public component is empirically different from the violent, militarized activities traditionally captured in the armed component. Likewise, the auxiliary component speaks to the population of noncombatant supporters who participate within the rebel organization but who do not openly indicate their sympathy or involvement. Civilians aiding a rebel organization either administratively and/or logistically, while simultaneously retaining their role as noncombatants, largely make up this aspect of an insurgency's operation. In contrast, the public component captures the public interfaces that rebel groups create to credibly interact with civilian populations that are not actively involved in the organization, be it openly or not.

The difference between the underground component as it is currently understood and the public component is the clandestine nature of the underground. By its very definition, the underground component operates almost exclusively in the shadows, out of the public eye. This characteristic is key to ensuring organizational security, particularly in urban environments where insurgents have little to no territorial control and battle strong states. Although in these conditions the underground component can fulfill some basic functions of a public component, such as disseminating propaganda or vetting new recruits, most of the activities associated with the public component require insurgents to engage in overt relations with the targeted public. As an example, the M-19 operated in major urban centers in Colombia. For most of its history, the M-19 relied heavily on its underground to maintain security and continue operations. The M-19 underground maintained a decentralized cellular structure with strict compartmentalization. When the insurgents met, all wore face masks, used noms de guerre, and took other measures to ensure that no one's identity was revealed. As a result, the M-19's ability to develop a robust public component was fairly limited. It was not until the group established "peace camps" as part of negotiations with the federal government that the M-19 used a public component. The peace camps were urban demilitarized zones where the M-19 could freely, and legally, engage with its target population. One insurgent described how the peace camps were one of the few places the guerrillas were able to take off their masks and speak with civilians in the neighborhood as M-19 members:

> We determined the needs of the people in those neighborhoods and then carried out an armed operation to hand out food, clothes, and even construction materials. The muchachos would perform this operation with their faces unmasked. We reasoned that clandestine operations were for the enemy, not the people. We had to be careful with our identity, but we would show our faces, so they could see us in flesh and blood solving their daily problems with them.[65]

The insurgents took part in numerous shadow governance activities while part of the peace camps, such as intervening on behalf of locals against exploitative tactics of landlords. As another example, the shadow governance activities of Hizbollah, some of the most robust in the world, are not undertaken by clandestine elements of the group's underground component. It would have been exceedingly difficult for the group to open hospitals and provide electricity, education, and garbage services to the residents in southern Beirut without doing so overtly and openly—garbage truck drivers and hospital workers do not wear balaclavas. In this regard, there is a qualitative distinction between the underground component and the public component.

The difficulty in explaining Hizbollah's social services wing as being an element of either its underground or armed component highlights a serious shortcoming within the current doctrine's capacity to map onto the empirical reality being observed, especially when considering that much of Hizbollah's support comes from its provision of social services to the Shia population in Lebanon. In this regard, expanding the heuristic model to account for these types of activities and strategies will offer the doctrine as a whole greater leverage when describing, contextualizing, and analyzing these sorts of phenomena.

The contextual factors that inevitably influence the scope and depth of an insurgency group's public component are largely a function of state strength. As noted above, the importance of territorial control in formulating and maintaining a robust public component is central to understanding why or why not rebel groups opt to create a public interface. Without the freedom of movement that territorial control allows, insurgent groups are often too hard-pressed to engage in shadow governance activities, much less engage in the legal processes of the state. Thus, territorial control is key in explaining the emergence of a public component, and as noted before, the ability to capture and hold territory diminishes greatly when pitted against a strong state.[66] On the other hand, if the insurgents are battling a weak state, the possibility of territorial control, and a robust public component, increases dramatically. For instance, in Lebanon, Hizbollah's success was predicated on a weak central Lebanese government. Its ability to secure territory in Beirut and the Bekaa Valley would have been severely hampered if the central government had the capacity to rebuff its control. Indeed, the Lebanese state had historically ignored the Shia populations, offering Hizbollah (and Amal) an opening to gain a strategic foothold in those areas.

The public component can offer insurgent organizations strategic advantages that the use of violence alone cannot. However, the utility of any one component—be it armed, underground, auxiliary, or public in nature—needs to be taken in context. Within the realm of irregular warfare, each component of an insurgency seeks to achieve specific ends. By understanding the strategic purpose of each component, one's conception of how these mechanisms are used in concert to achieve victory becomes clearer. With that said, the formation of a public component should be read in a positive light. The formation and use of a public component signals that a rebel group may have begun to move away from a campaign of absolute victory; in addition, the existence of a public interface increases the possibility of a negotiated settlement or power-sharing agreement. Since the end of the Cold War, more and more political conflicts are being resolved via negotiated settlements. The prevalence of negotiation as credible exit to a conflict highlights how victory is no longer perceived in absolute terms but rather as a series of conditions. The activities inherent to the public component are crucial in extending an insurgency group's bargaining space, as violence alone cannot deliver concessions. Furthermore, it requires that the rebel groups interface with the public in order to demonstrate their legitimacy and capacity to rule. Further investigation into the variety of ways violent nonstate groups can potentially use the public component to gain power and resources, especially in weak or failing states, is needed; however, the above analysis has taken the first step in this direction. By articulating the strategic nature that the public component serves within irregular war environments, a more precise picture of modern conflicts emerges.

NOTES

[1] Mark Lichbach, *The Rebel's Dilemma* (Michigan: University of Michigan Press, 1995), 18.

[2] Nathan Bos, ed., *Human Factors Considerations of Undergrounds in Insurgencies* (Fort Bragg, NC: United States Army Special Operations Command, 2013), 35–36.

[3] Ibid.

[4] Lotta Themnér and Peter Wallensteen, "Armed Conflicts, 1946–2011," *Journal of Peace Research* 49, no. 4 (2012): 565–575.

[5] Monica Duffy Toft, *Securing the Peace: The Durable Settlement of Civil Wars* (Princeton, NJ: Princeton University Press, 2010).

[6] Ibid.

[7] Joakim Kreutz, "How and When Armed Conflicts End: Introducing the UCDP Conflict Termination Dataset," *Journal of Peace Research* 47, no. 2 (2010): 243–250.

[8] Human Security Centre, *Human Security Report 2005: War and Peace in the 21st Century* (New York: Oxford University Press, 2005).

[9] Mary Kaldor, *New and Old Wars: Organized Violence in a Global Era* (Cambridge, England: Polity Press, 2012).

[10] Luis de la Calle and Ignacio Sanchez-Cuenca, "Rebels without a Territory: An Analysis of Nonterritorial Conflicts in the World, 1970–1997," *Journal of Conflict Resolution* 56, no. 4 (2012): 583.

[11] Ami Pedahzur and Leonard Weinberg, *Political Parties and Terrorist Groups* (New York: Routledge, 2013).

[12] Menapolis, "Local Councils in Syria: A Sovereignty Crisis in Liberated Areas," policy paper (Istanbul: Menapolis, September 2013).

[13] Maram Mazen, "Bloodshed Corrodes Support for Boko Haram," *Al Jazeera*, May 25, 2014.

[14] Liz Philipson and Yuvi Thangarajah, *The Politics of the North-East: Part of the Sri Lanka Strategic Conflict Assessment 2005* (Colombo, Sri Lanka: The Asia Foundation, 2005).

[15] Summer Newton and Robert Leonhard, "Shadow Government," in *Undergrounds in Insurgent, Revolutionary, and Resistance Warfare,* ed. Robert Leonhard (Fort Bragg, NC: United States Army Special Operations Command, 2013), 133–134.

[16] Y. Hasenfeld, "Power in Social Work Practice," *Social Service Review* 61, no. 3 (1987): 469–483.

[17] Stathis N. Kalyvas, *The Logic of Violence in Civil War* (New York: Cambridge University Press, March 2006), 87–89.

[18] Mahmood Mamdani, *When Victims Become Killers* (Princeton, NJ: Princeton University Press, 2001), 186–189.

19 Shawn Teresa Flanigan and Mounah Abdel-Samad, "Hezbollah's Social Jihad: Nonprofits as Resistance Organizations," *Middle East Policy* 16, no. 2 (June 2009): 122–137.

20 Kalyvas, *The Logic of Violence in Civil War*, 114.

21 Nelson Kasfir, "Guerrillas and Civilian Participation: the National Resistance Army in Uganda, 1981–86," *Journal of Modern African Studies* 43, no. 2 (2005): 271–296.

22 Zachariah Cherian Mampilly, *Rebel Rulers: Insurgent Governance and Civilian Life during War* (Ithaca, NY: Cornell University Press, 2011), 54.

23 Alberto Bolivar, "Peru," in *Combating Terrorism: Strategies of Ten Countries*, ed. Y. Alexander (Ann Arbor: University of Michigan Press, 2005), 84–115.

24 Shawn Teresa Flanigan, "Charity as Resistance: Connections between Charity, Contentious Politics, and Terror," *Studies in Conflict & Terrorism* 29, no. 7 (2006): 641–655.

25 For more, see Sarah Elizabeth Parkinson, "Organizing Rebellion: Rethinking High-Risk Mobilization and Social Networks in War," *American Political Science Review* 107, no. 3 (2013): 418–432; and Robert Leonhard, "Recruiting" in *Undergrounds in Insurgent, Revolutionary, and Resistance Warfare*, ed. Robert Leonhard (Fort Bragg, NC: United States Army Special Operations Command, 2013), 19–42.

26 Shimon Shapira, *Hezbollah between Iran and Lebanon* (Tel Aviv: Hakibbutz Hameuchad, 2000), 140.

27 Ahmad Nizar Hamzeh, *In the Path of Hizbullah* (Syracuse: Syracuse University Press, 2004).

28 Eyal Zisser, "Hizballah in Lebanon: At the Crossroads," *MERIA* 1, no. 3 (1997).

29 Shawn Teresa Flanigan, "Nonprofit Service Provision by Insurgent Organizations: the Cases of Hizballah and the Tamil Tigers," *Studies in Conflict & Terrorism* 31, no. 6 (2008): 499–519.

30 Mampilly, *Rebel Rulers*.

31 Philipson and Thangarajah, *The Politics of the North-East*.

32 Flanigan, "Nonprofit Service Provision by Insurgent Organizations," 511.

33 Philipson and Thangaraiah, *The Politics of the North-East*.

34 Ibid.

35 A. J. V. Chandrakanthan, "Eelam Tamil Nationalism: An Inside View," in *Sri Lankan Tamil Nationalism: Its Origins and Development in the Nineteenth and Twentieth Centuries*, ed. A. Jeyaratnam Wilson (Vancouver: University of British Columbia Press, 2000).

36 Mampilly, *Rebel Rulers*, 190–206.

37 Shaul Mishal and Avraham Sela, *The Hamas Wind: Violence and Compromise* (Tel Aviv: Miskal, 2006), 195–196.

38 Steven Wilkinson, *Votes and Violent: Electoral Competition and Ethnic Riots in India* (New York: Cambridge University Press, 2006).

39 Pedahzur and Weinberg, *Political Parties and Terrorist Groups*, 72.

40 Patrick Bishop and Eamonn Mallie, *The Provisional IRA* (London: Corgi, 1992), 39.

41 Michael Noonan, "Ahern Says His View of Sinn Fein Remains the Same," *Irish Times*, May 3, 2001.

42 A. Richards, "Terrorist Groups and Political Fronts: the IRA, Sinn Fein, the Peace Process and Democracy," *Terrorism and Political Violence* 13, no. 4 (2001): 72–89.

43 Ed Moloney, *A Secret History of the IRA* (New York: W. W. Norton, 2002).

44 Pedahzur and Weinberg, *Political Parties and Terrorist Groups*, 132.

45 Adrian Guelke and Jim Smyth, "The Ballot Bomb: Terrorism and the Electoral Process in Northern Ireland," in *Political Parties and Terrorist Groups*, ed. Leonard Weinberg (London: Frank Cass, 1992), 103–124.

46 George Mitchell, *Making Peace* (New York: Alfred Knopf, 1999), 23–24.

47 Pedahzur and Weinberg, *Political Parties and Terrorist Groups*, 134.

48 Richards, "Terrorist Groups and Political Fronts," 75.

49 Avraham Sela, *Non-State Peace Spoilers and the Middle East Peace Efforts* (Jerusalem: The Floersheimer Institute for Policy Studies, 2005).

50 Arnon Reguler, "Achievement to Hamas in Local Elections: Won Most of Big Municipalities," *Haaretz*, May 8, 2005.

51 M. Van Engeland Anisseh and R. Rachael, *From Terrorism to Politics* (Vermont: Ashgate, 2007).

52 Ibid., 19.

53 Pedahzur and Weinberg, *Political Parties and Terrorist Groups*, 2013.

54 Norton Augustus Richard, *Hezbollah: A Short History* (Princeton, NJ: Princeton University Press, 2007), 101.

55 Magnus Ranstorp, "Hezbollah's Command Leadership: Its Structure, Decision-making and Relationship with Iranian Clergy and Institutions," *Terrorism and Political Violence* 6, no. 3 (1994): 303–339.

56 Krista E. Wiegand, "Reformation of a Terrorist Group: Hezbollah as a Lebanese Political Party," *Studies in Conflict & Terrorism* 32, no. 8 (2009): 669–680.

57 Gulistan Gurbey, "The Kurdish Nationalist Movement in Turkey Since the 1980s," in *The Kurdish National Movement in the 1990s*, ed. Robert Olson (Lexington, KY: The University Press of Kentucky, 2006), 15.

58 Philip Robins, "The Overlord State: Turkish Policy and the Kurdish Issue," *International Affairs* 69, no. 4 (1993): 661.

59 Ayla Yackley, "Turkey Outlaws Kurdish Political Party," *The Globe and Mail*, 2003.

60 David Bushnell, *The Making of Modern Colombia* (Los Angeles, CA: University of California Press, 1993), 201–205.

61 Pedahzur and Weinberg, *Political Parties and Terrorist Groups*, 132.

62 Jeremy M. Weinstein, *Inside Rebellion the Politics of Insurgent Violence* (New York: Cambridge University Press, 2007), 27–60.

63 Matthew Levitt, "Hamas from Cradle to Grave," *Middle East Quarterly* 11, no. 1 (2004): 1–12. Also see Flanigan, "Nonprofit Service Provision by Insurgent Organizations."

64 Eli Berman and David D. Laitin, "Religion, Terrorism and Public Goods: Testing the Club Model," *Journal of Public Economics* 92, nos. 10–11 (2008): 1942–1967.

65 María Eugenia Vásquez Perdoma, *My Life as a Colombian Revolutionary: Reflections of a Former Guerillera* (Philadelphia: Temple University Press, 2005), 194.

66 de la Calle and Sanchez-Cuenca, "Rebels without a Territory."

INTEREST, IDENTIFICATION, INDOCTRINATION, AND MOBILIZATION: I³M

Jesse Kirkpatrick

INTRODUCTION

In 2013, researchers from the Johns Hopkins University Applied Physics Laboratory (JHU/APL) and United States Army Special Operations Command (USASOC), G3 Sensitive Activities Division, attempted to delineate the factors leading up to and including insurgent mobilization. The result of this collaborative effort was the development of a new model of insurgent participation and involvement that leverages the findings of social science research while focusing on the four critical areas of interest, identification, indoctrination, and mobilization (I^3M).[a] I^3M is a model that at once draws upon the existing literature and, in some cases, simplifies and streamlines existing models, while also attempting to correct for some deficiencies found in the contemporary insurgency scholarship.

This thought piece explores how and why individuals become insurgents, with the ultimate goal of further developing the framework and boundaries of the conceptual I^3M model by using existing empirical data, social science literature, and case studies. Drawing from the I^3M model, this piece examines the process individuals go through to become active members of insurgent organizations by focusing on four phases: interest, identification, indoctrination, and mobilization (hereafter, I^3M).

This piece proceeds in four parts. Part one introduces the I^3M model while situating it within the current scholarship on insurgent participation and motivation. Part two uses a brief case study to explore and demonstrate how I^3M may be used to understand an individual's progression toward becoming an insurgent. Part three concludes.

WHY MEN AND WOMEN FIGHT:
A LAY OF THE LAND

Early scholarship on the genesis of political violence, of which insurgency is a part, was typically produced by historians who tended to focus largely on particular instances and cases of violence, as opposed to general theories of its origin.[2,b] Consequently, political scientists sought to take the particularized insights of historians and apply them

---

[a] The original developers of the I^3M model are Bruce DeFeyter and Christina Phillips, with subsequent developmental contributions made by Jesse Kirkpatrick and Mary Kate Schneider. For a brief overview of the I^3M model, see below; for a more thorough account, see Kirkpatrick and Schneider.[1]

[b] Political violence is taken to mean violence that is carried out with the aim of effecting political change or maintaining the political status quo. The term encompasses, but is not limited to, insurgencies and terrorism.

toward developing generalizable theories of the origin of political violence. In turn, for much of the 1990s, social scientists focused on the origins of large-scale interstate wars. One result of this effort was that it "tended to crowd out the study of other arenas for political violence as well as other forms of violence, including civil wars, genocide (and other types of mass killing), ethnic cleansing, and terrorism."[3] The outcome of such "crowding out" is that only in the very recent past have political scientists turned their attention to studying these other forms of political violence, such as insurgencies.

Although the widening in the study of political violence was certainly a welcome development, shortcomings nevertheless remained. In what one political scientist terms the "pre-1990s consensus," scholars characterized participation in political violence (in this case, political violence leveled against noncombatants) as irrational, random, and motivated by ancient, primordial hatreds.[4] In fact, this so-called consensus endured throughout the 1990s and was typified in such pieces as Kaplan's *Balkan Ghosts*[5] (in which he argues that the Balkans were torn apart because of ancient hatreds between various religious groups) and Huntington's *Clash of Civilizations*[6] (in which conflict is viewed as originating from incompatibilities between civilizations). By characterizing such political violence as illogical, wanton, and senseless, these analyses tended to exclude the possibility that participants in political violence may, in some cases, be subject to and involved in more systematic and rational processes.

A notable exception to the pre-1990s and 1990s consensus came from a small group of scholars whose focus was terrorism.[7,c] These scholars rejected the theses that participation in political violence is rooted in illogical, gut-level, deep-seated, ancient hatred, instead taking as their point of departure the possibility that such violence could be logical, strategic, purposive, and motivated and embarked upon by rational actors who are animated by reasons and incentives other than those related to mere irrationality.

As political scientists began to expand their cases of study to include insurgencies, revolutions, civil wars, and terrorism, they sought to explain not just the general causes for these types of conflict but also the motivations behind the various actors engaged in them.

---

[c] Within this scholarship exist many differing definitions of *terrorism*. For the purposes of this piece, and following Title 18, Part I, Chapter 113B, § 2331 of the United States Code, *terrorism* is taken to mean acts that are intended "to intimidate or coerce a civilian population; (ii) to influence the policy of a government by intimidation or coercion; or (iii) to affect the conduct of a government by mass destruction, assassination, or kidnapping."

While disagreement endures, scholars have pointed to various exogenous causes or risk contributors for conflict. These include socio-economic and political conditions such as poverty,[8] income disparity and material depravation,[d] and political exclusion that may give way to endogenous motivations and incentives for individuals' participation in conflict. For example, poverty may reduce individuals' opportunity costs, thus making them more susceptible to identification and recruitment efforts.[10]

Since September 11, 2011, many continued this research and sought to puzzle out just why it is that men and women fight, why terrorists become terrorists, and why insurgents become insurgents.[11,e] Many of these scholars developed typologies and models that attempt to identify and explicate external conditions, motivations, and incentives that may cause or act as contributing factors toward individuals' involvement in insurgent activity.

Much of this recent scholarship has focused on the process of radicalization as a key component for insurgent motivation.[14] While it is certainly true that *some* insurgents are radicals, not all insurgents are radicalized.[f] Insurgents may be motivated by factors that have little or nothing to do with root or fundamental change—a key component of radicalization. Consequently, this area of research excludes a large number of insurgencies, thus failing to explore the incentives and motivations of nonradicalized insurgents. Therefore, one of the ultimate goals of the JHU/APL–USASOC research team was to account for such deficiencies and create a model that is objective and normatively neutral and may be applied across a range of insurgency cases.

---

[d] Income disparity may lead to relative depravation, the feeling that one is entitled to a good that one lacks while others in the population possess or have access to said good(s).[9]

[e] Note the distinction between *terrorism* and *insurgency*. The former, by definition (see note c), requires the intent to intimidate or coerce a civilian population, whereas the latter does not require the use of civilians (although it can). This is not to say that insurgents cannot engage in terrorism and become terrorists. For a similar definition of insurgency, see the *U.S. Government Counterinsurgency Guide*, which defines insurgency as "the organized use of subversion and violence to seize, nullify, or challenge political control of a region."[12] Also see the *Casebook on Insurgency and Revolutionary Warfare, Volume II: 1962–2009*, which defines insurgency as "an attempt to modify the existing political system at least partially through unconstitutional or illegal use of force or protest."[13]

[f] *Radical* is here defined as "associated with political views, practices, and policies of extreme change."[15]

I³M: AN OVERVIEW

Part One of the Model

I³M is a two-part model. The first part captures four behavioral processes. These include interest (what piques the curiosity of new recruits?); identification (what leads new recruits to associate them-selves with the movement?); indoctrination (what causes new recruits to take the leap and actually join the movement?); and mobilization (the point at which individuals take action in support of the movement). Interest, identification, and indoctrination are the three factors lead-ing up to mobilization.

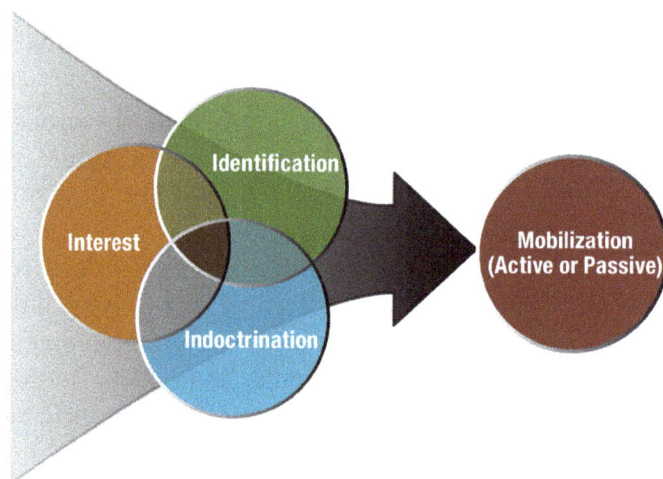

Figure 1. Four behavioral processes of I³M.

It is important to note that while the components of I³M are inter-connected, the processes are not determinatively linear; nor is there a single explanation for why individuals decide to mobilize.[g] While some individuals may be sympathetic to or in agreement with the message, aim, and tactics of an insurgency, it is no secret that most individuals will not be moved to action.[h] This raises the perennially interesting question: what, then, motivates or incentivizes individuals to become involved in an insurgency?

---

[g] As a developer of I³M, I agree with Patel that it is incorrect to argue that "contrary to empirical social science studies . . . the path to terrorism has a fixed trajectory and that each step of the process has specific, identifiable markers."[16]

[h] As an example, we can call to mind that approximately less than 1 percent of the US population serves in the US military, a number certainly dwarfed by those in the United States who identify with or have a sympathetic interest in the activities and mission of the armed forces.

Part Two of the Model: Motivations and Incentives

Here we can turn to the second part of the I³M model, which identifies some general motivations and incentives for why individuals may engage in the I³M process. Following is an outline of nine representative examples of motivations and incentives that are captured in the I³M model. In identifying these motivations and incentives, the team drew from the aforementioned empirical work of insurgency scholars. While the sample populations that these scholars focus on varies from radicals in Saudi Arabia to secular combatants in Africa, an appropriate usage of these scholars' typologies and models can help illuminate the contributing factors to insurgent interest and identification (a key part of motivation). This, in turn, can be used to help explain the process through which individuals become indoctrinated and eventually mobilized.[17]

Motivations and Incentives

1. **Personal grievances:** Grievance resulting from personal experience, such as observed or perceived injustices (e.g., torture of a family member) or relative depravation.[18]

2. **Group grievances:** Grievances resulting from the treatment or perceived treatment of a collective, i.e., group (e.g., one's conationals or members of one's religious community).

3. **Slippery slope:** Small instances and cases of, and exposure to, insurgent violence may incrementally increase one's tolerance, acceptance, or endorsement of such activity.

4. **Love or affection:** Whether it is having a spouse or lover in a group or the desire for "romantic or comradely love," love can be a strong motivator for insurgent I³M.[20]

5. **Risk and status:** Some individuals may find the status that comes with insurgent membership attractive. Likewise, the draw of risk taking may also be appealing.

6. **Unfreezing:** Some individuals value the camaraderie and social capital that can result from participation in an insurgency; insurgent groups and their members can sometimes fill the need for companionship.

7. **Group polarization:** The polarization of a group of which one is a member. A common example of group polarization is so-called prison universities, or the imprisonment of individuals who have existing affiliations, sympathies, and allegiances to a given insurgent group or those who are susceptible to creating such an affiliation.[i]

8. **Group competition:** Competition between groups, be it political, economic, or social, can sometimes lead individuals to engage in I^{3}M.

9. **Group isolation:** The extant or perceived isolation or repression of a group may lead individuals in the group to engage in increased mobilization.

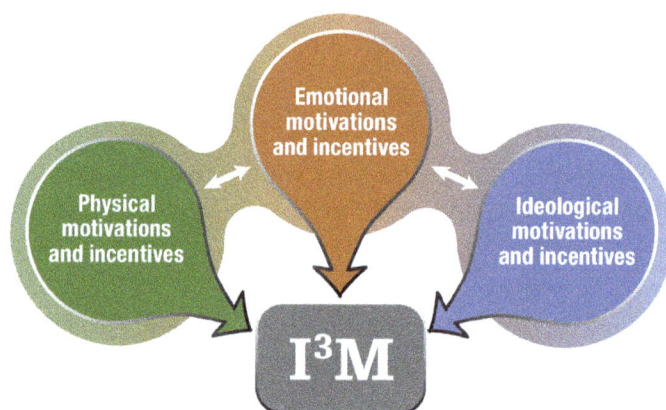

Figure 2. Motivations for I^{3}M.

In an attempt to streamline and simplify these motivations and incentives, the I^{3}M model divides them (as depicted in Figure 2) into three categories: physical, emotional, and ideological.[j] It is important to note that although the tripartite division of these categories is a useful heuristic, in reality, the motivations and incentives influencing I^{3}M can be cross-categorized. For example, imagine that a neo-Nazi Caucasian male residing in the United States witnessed law enforcement officers subdue his father through the use of force. This action,

---

[i] It should be noted that with changes to communication and technology, groups can now be virtual (e.g., online), with a strong shared identity despite having never met in person.

[j] Some scholars have disaggregated "mechanisms" of political radicalization on the basis of the number of individuals involved, distinguishing between the individual, group, and mass levels.[21] While some of these mechanisms may correspond with I^{3}M's individual motivations and incentives, the I^{3}M model applies to the individual level. For a defense of this approach, see Hegghammer.[22] The current approach does not, of course, preclude further I^{3}M research examining motivations and incentives at these larger levels.

consequently, results in the son developing a personal grievance against law enforcement writ large. It is perfectly plausible that the motivations arising from this grievance could be categorized as physical (his desire to safeguard his and his father's physical safety), emotional (the love he has for his father and his desire for redress of his grievances), and ideological (the belief that he and his family have a right to engage in the unlawful activity for which his father was arrested). It is in this way that the categorizations outlined in the I³M model are, admittedly, not entirely separable.

Similarly, individuals' motivations and incentives for I³M are pluralistic. Returning to our example, the son may be interested in an insurgency, say, the neo-Nazi group White Knights United, for a variety of reasons. Perhaps it is because of his strong religious and political beliefs, but perhaps there are also enormous social pressures.[k] Just like in everyday life, individuals are influenced by numerous motives, which are further impacted by additional factors such as environment, social conditions, life history, and so forth.

CASE STUDY: AUTODEFENSAS UNIDAS DE COLOMBIA[24]

This brief section explores the short case study of the United Self-Defense Forces of Colombia (Autodefensas Unidas de Colombia, or AUC) in the hopes of explicating some of the operative incentives and motivations that may contribute to an individual's transformation into an insurgent, with the anticipation that future research may more fully explore potentially recognizable indicators of when an individual is progressing toward mobilization in an insurgency.

Background

Formed in 1997 and officially demobilized in 2006, the AUC was the main right-wing insurgency group in Colombia. The group served as the umbrella organization for the various semiautonomous paramilitaries operating in Colombia during its decades-long civil war. The group's historical legacy and origination dates back to the early 1950s, a period when wealthy landowners, ranchers, and businesspeople sought to protect their property and physical security against leftist insurgents. In doing so, these stakeholders backed and armed various right-wing militias. Later, finding a common enemy in the leftist

---

[k] For an interesting computational prototype of a quantitative model that explores, in part, the role of social pressures on behavior, see Davis and O'Mahoney.[23]

guerrillas, the Colombian political and military establishment sought to collaborate with, support, and, ultimately, give legal protection to these "self-defense" groups.

In the 1980s, at the height of Colombia's booming narcotics trade, the groups served as protectors of Colombia's drug lords. By the 1990s, Colombia's narco-kingpins' power waned, and the self-defense forces eventually moved into direct involvement with narcotrafficking, cementing their positions as the country's leading narcotraffickers. The staple tactic of the AUC was violence; targets included noncombatant peasants, leftist guerrillas, politicians, and, in some cases, the Colombian military. This use of ruthless violence, such as the targeting of civilians, massacre of peasants, and attacks on the military, resulted in the group's designation as a terrorist organization by the United States on September 10, 2001.[1]

Early in the peace talks between the Fuerzas Armadas Revolucionarias de Colombia (the FARC) and the Colombian government, the FARC sought political legitimacy and recognition. Evidence indicates that this strategy was, in fact, pursued prior to disarmament negotiations; as early as 1991, Fidel Castaño advocated that the AUC's groups "construct a solid and coherent political platform."[27] This strategy followed the AUC's narrative and ideology of acting as a steward and protector of the middle class and landed elite against left-wing guerrillas. The AUC could at once engage in violence while claiming that it was doing so not to enrich its members but to act as a bulwark against the looming guerrilla threat.

In addition to aiding the group's strategic communication efforts, the quest for legitimacy had a perhaps more important outcome for the group: knowing that when the government concluded a peace agreement with the left-wing guerrillas they would be without peer, the AUC understood that its size and power would ensure it a strong negotiating position and favorable disarmament conditions. This strategy resulted, in part, from the Colombian government's disparate treatment of what it characterized as mere criminal narcotraffickers in contrast to organized armed rebel groups.[m] The former were pursued as states often pursue criminals—with arrest and prosecution—while the latter were treated as actors for which such a strategy was ineffective. Consequently, like many counterinsurgencies, the group was militarily eroded while offers of talks and favorable conditions were mounting.

---

[1] For a statistical compilation of the dates, actors, types, and locations of this violence, see the Global Terrorism Database.[25]

[m] For a discussion of how drug lords used the state's treatment of the AUC, in contrast to its treatment of run-of-the-mill "criminal" groups, to their advantage, namely by folding (typically for a large fee) their drug outfits into the AUC, see Hristov.[26]

As domestic and international pressure began to grow for Colombia to end its internal armed disputes, the government and the AUC entered into negotiations. Consequently, in 2003, under the Santa Fe de Ralito Accord, the group began demobilization. Although the demobilization was largely successful, reports suggest that remnants of the organization have reformed and continue to engage in violence and illegal activity.[n]

Motivations and Incentives for Members

The clandestine and opaque nature of the AUC makes it difficult to ascertain motivations for why members were interested in, identified with, were indoctrinated into, and, ultimately, mobilized for the AUC. In addition, one must appreciate the variation not only from individual to individual but also between the leadership and the lower rank-and-file members. Despite these caveats, the AUC leadership has maintained a public enough presence (through sources such as media interviews, court appearances, intelligence analysis, and their own online web presence) that one can arrive at some relatively firm conclusions regarding members' motivations for joining and remaining in the AUC.

Although the AUC defined itself "as an anticommunist advance guard in 'defense of private property and free enterprise,'" thus advancing a grievance narrative in which hardworking business owners, adherents of "traditional values," and nonleftists were under the threat of the guerrillas, leaders were motivated largely by military power that permitted them to acquire large amounts of wealth.[28] Throughout the group's existence, AUC leaders sought to secure arms, land, political allies, and armed members for the purpose of enriching themselves economically. According to Chernick, the AUC's leadership was most interested in land—its use, its acquisition, and keeping it in the hands of rural oligarchs.[29] Such land policies ensured the AUC control over illicit drug production and smuggling routes, as well as direct links to petroleum pipelines and areas of resource extraction.

An interesting piece of the group's history involves its leaders, Carlos and Fidel Castaño. The Castaño brothers' father was murdered in 1981, leading the brothers to establish the right-wing self-defense group Las Tangas. It would seem, then, that their early involvement in the insurgency was motivated by a mix of personal grievance, love, and group grievance (under the notion that their traditional way of life was

---

[n] Precise data and information about these attacks is difficult to obtain, largely because many of these groups have been renamed and operate in the shadows.

under attack by the leftists), which can be categorized as physical, emotional, and ideological, respectively.

Perhaps more interesting than the motivations of the small number of top AUC leaders are those of the rank-and-file members. We are lucky to have two sets of in-depth interviews that scholars conducted with demobilized AUC members.[30] These studies suggest that

> with ex-combatants of the AUC, their principal reasons for joining were: via an acquaintance who convinced the person to join (29 percent); because they lived in a zone under paramilitary control and joining was "just what you did" (17 percent); recruited by force or threat (14 percent); or economic motivations (27 percent).[31]

This research underscores two important possibilities: first, although some members may have been radicalized, this sample pool of interviewees indicates that most were not; and second, familial, peer, and social networks played an influential role in motivating individuals to join the AUC. Many members were also motivated by a desire for revenge, "a love of arms," and the hope of increasing their economic opportunities.[32] Another ex-combatant cited boredom, social status, and good pay.[33]

In addition to these motivations, social expectations and a culture of paramilitaristic machismo and status played a role in individuals joining the AUC and remaining in the group. Interviewees cite membership in the AUC as providing a feeling of self-worth, along with incentives such as respect from those in their communities, the ability to obtain fine clothes, and the lure of beautiful women and love.[34]

Retention (i.e., continued mobilization) was ensured by the AUC's pay structure and incentives to kill. Members in command positions received a significant increase in salary and also bonuses tied to the number of guerrillas they killed and fought.[35] As one ex-combatant put it, commanders had "cars, motorcycle, guns, and staff" that made their lives easier.[36] Leadership positions were so coveted, and competition was so stiff that AUC members are reported to have killed family and friends just to ascend the ranks.[37,o]

---

o Again, interestingly, these motivations and incentives do not seem to indicate radicalization of most members.

CONCLUSION: QUESTIONS AND SUGGESTIONS FOR FURTHER STUDY

Although no model will be perfectly predictive of what will motivate and incentivize a person to engage in an insurgency, I³M offers the promise of a streamlined yet robust analytical tool. The model harnesses existing findings in the scholarship in a way that does not foreclose future refinements, adjustments, and revisions to I³M.

This thought piece is intended to serve as an in-depth overview of the I³M model, a demonstration of its efficacy in application by way of case study, and, finally, a call for future study. The study and practice of (counter)insurgency could benefit from developing the I³M model. It is the hope of this author that future studies of insurgency will take I³M seriously.

Such future study of I³M could include an in-depth empirical study not dissimilar to the small case study found in this piece. The cases could be appropriately selected to ensure variation across types of insurgencies, when and where they occurred, and their participants. This increase and variation in cases could help tease out, identify, and in some cases, eliminate our currently speculative conclusions regarding insurgent motivations and incentives. A thorough analysis of the Assessing Revolutionary and Insurgent Strategies casebooks would be a good place to start.

In addition to an increase in the number and types of cases under analysis, I³M would also benefit from a real-time application. This should include analyses of current insurgencies such as the conflicts in Syria and Iraq. Open sources such as YouTube and Twitter, and the insurgents' active, daily use of these technologies, provide a peek into the mediated, public reasons for why individuals are interested in an insurgency, why they identify with it, why they mobilize, and, although in far fewer cases, why they leave. In addition, further study should include conflicts such as those in Afghanistan that are (potentially) winding down and for which there exist numerous and extensive interview records.

These are just a few examples of how I³M could be developed, tested, and put to use.

NOTES

1 Jesse Kirkpatrick and Mary Kate Schneider, "I³M–Interest, Identification, Indoctrination, and Mobilization: A Short Introduction to a New Model of Insurgent Involvement," *Special Warfare Magazine* 26, no. 4 (October–December 2013): 23–27.

2 For a review of these developments, see Benjamin A. Valentino, "Why We Kill: The Political Science of Political Violence against Civilians," *Annual Review of Political Science* 17 (2014): 89–103, 90.

3 Ibid., 91.

4 Ibid.

5 Robert D. Kaplan, *Balkan Ghosts: A Journey through History* (New York: St. Martin's, 1993).

6 Samuel P. Huntington, "The Clash of Civilizations?" *Foreign Affairs* 72, no. 3 (1993): 22–49.

7 See, for example, Martha Crenshaw, "The Causes of Terrorism," *Comparative Politics*, 13 no. 4 (July 1981): 379–399; and B. M. Jenkins, "International Terrorism: A New Mode of Conflict," in *International Terrorism and World Security*, ed. D. Carlton and C. Schaerf (London: Croom Helm, 1975), 13–49. For a representative sample of the early debates in the 1990s, see Walter Reich, *Origins of Terrorism: Psychologies, Ideologies, Theologies, States of Mind* (1998; repr., Washington, DC: Woodrow Wilson Center, 1990).

8 Nathan Bos, ed., *Human Factors Considerations of Undergrounds in Insurgencies* (Fort Bragg, NC: United States Army Special Operations Command, 2013), 15.

9 Ted Robert Gurr, *Why Men Rebel* (Princeton, NJ: Princeton University Press, 1970).

10 Paul Collier and Anke Hoeffler, "Greed and Grievance in Civil War," *Oxford Economic Papers* 56, no. 4 (2004): 563.

11 See Crenshaw, "The Causes of Terrorism," 379–399; Bruce Hoffman, *Inside Terrorism* (New York: Columbia University Press, 1998); Eli Berman, *Radical, Religious, and Right: The New Economics of Terrorism* (Cambridge, MA: Massachusetts Institute of Technology Press, 2009); Thomas Hegghammer, "Terrorist Recruitment and Radicalization in Saudi Arabia," *Middle East Policy*, 8, no. 4 (2006): 39–60; James W. Jones, *Blood That Cries out from the Earth: The Psychology of Religious Terrorism* (Oxford: Oxford University Press, 2008); Clark McCauley and Sophia Moskalenko, *Friction: How Radicalization Happens to Them and Us* (Oxford: Oxford University Press, 2011); Clark McCauley and Sophia Moskalenko, "Mechanisms of Political Radicalization: Pathways toward Terrorism," *Terrorism and Political Violence*, 20, no. 3 (2008): 415–433; Faiza Patel, *Rethinking Radicalization* (Brennan Center for Justice, New York University School of Law, 2011); and Ronald Wintrobe, *Rational Extremism: The Political Economy of Radicalism* (Cambridge: Cambridge University Press, 2006).

12 U.S. Government Interagency Counterinsurgency Initiative, *U.S. Government Counterinsurgency Guide* (Washington, DC: Bureau of Political-Military Affairs, Department of State, 2009).

13 *Casebook on Insurgency and Revolutionary Warfare, Volume II: 1962–2009*, ed. Chuck Crossett (Fort Bragg, NC: United States Army Special Operations Command, 2009), xvi.

14 Hegghammer, "Terrorist Recruitment and Radicalization," 39–60; McCauley and Moskalenko, *Friction*; and McCauley and Moskalenko, "Mechanisms of Political Radicalization."

15 *Radical* as defined by Merriam-Webster.com, http://www.merriam-webster.com/dictionary/radical.

16 Patel, *Rethinking Radicalization*, 1.

17 For a discussion of the various kinds of mobilization and participation, see *Casebook on Insurgency and Revolutionary Warfare, Volume II: 1962–2009.*

18 See Syed Mansoob Murshed and Sara Pavan, *Identity and Islamic Radicalization in Western Europe: Economics of Security Working Paper 14* (Berlin: Economics of Security, 2009), 1–35, 4; for a discussion of relative depravation, see Gurr, *Why Men Rebel.*

19 See Stanley Milgram, *Obedience to Authority* (Harper Perennial Modern Classics, 2009).

20 McCauley and Moskalenko, *Friction*, 54.

21 Ibid.

22 Hegghammer, "Terrorist Recruitment and Radicalization in Saudi Arabia."

23 Paul K. Davis and Angela O'Mahoney, *A Computational Model of Public Support for Insurgency and Terrorism: A Prototype for More-General Social-Science Modeling* (Santa Monica, CA: RAND Corporation, 2013), http://www.rand.org/content/dam/rand/pubs/technical_reports/TR1200/TR1220/RAND_TR1220.pdf.

24 For a comprehensive review and analysis of the main right-wing insurgent group in Colombia, see Katharine Raley Burnett et al., "Autodefensas Unidas de Colombia (AUC)," in *Case Studies in Insurgency and Revolutionary Warfare—Colombia (1964–2009)*, ed. Summer Newton (Fort Bragg, NC: United States Army Special Operations Command, forthcoming), 269–319.

25 Global Terrorism Database, http://www.start.umd.edu/gtd/.

26 Jasmin Hristov, *Blood and Capital: The Paramilitarization of Colombia* (Athens, OH: Ohio University Press, 2009), 70.

27 James F. Rochlin, *Vanguard Revolutionaries in Latin America: Peru, Colombia, Mexico* (Boulder, CO: Lynne Rienner Publishers, 2003), 148.

28 Mauricio Romero, "Changing Identities and Contested Settings: Regional Elites and the Paramilitaries in Colombia," *International Journal of Politics, Culture and Society* 14, no. 1 (2000): 66.

29 Mark Chernick, "PCP-SL: Partido Communista de Peru-Sendero Luminoso," in *Terror, Insurgency, and the State: Ending Protracted Conflicts*, ed. Marianne Heiberg, Brendan O'Leary, and John Tirman (Philadelphia: University of Pennsylvania Press, 2007), 58–59.

30 For a fuller account of these interviews, their conclusions, and the motivation and behavior of AUC members, see Burnett et al., "Autodefensas Unidas de Colombia (AUC)," 303–305; and Cristina K. Theidon, "Transitional Subjects: The Disarmament, Demobilization and Reintegration of Former Combatants in Colombia," *International Journal of Transitional Justice* 1, no. 1 (2007): 66–90. Also see Universidad Nacional de Colombia, Observatorio de Procesos de Desarme, Desmovilización y Reintegración, *Dinámicas de las Autodefensas Unidas de Colombia* (Bogotá, August 2009), 16, http://alfresco.uclouvain.be/alfresco/d/d/workspace/SpacesStore/f6c1ffc2-2a02-4823-94fa-2de37c2a7bf1/Universidad%20Nacional%20-%20Dinamicas%20de%20las%20AUC%202009.pdf.

31 Theidon, "Transitional Subjects," 75.

32 Universidad Nacional de Colombia, *Observatorio de Procesos de Desarme, Desmovilización y Reintegración*, 16.

33 Theidon, "Transitional Subjects," 75–76.

34 Ibid, 76.

35 Universidad Nacional de Colombia, *Observatorio de Procesos de Desarme, Desmovilización y Reintegración*, 13.

36 Ibid.

37 Ibid.

THRESHOLD OF VIOLENCE

Guillermo Pinczuk and Jameel Khan

INTRODUCTION

Violence, at its core, is the deliberate infliction of harm on other people or things.[1] Within the arena of revolutionary warfare, actors use violence as a means to further their cause and realize political ends. Members of resistance or insurgent groups are those actors who seek to challenge, disrupt, or overthrow the existing order in a society, region, or country. Chief among the insurgent group's means to achieving this end is the use of violence. Both the insurgency and the government often employ various forms of violence with the ultimate objective of defeating the other side and cementing their own governance.

This paper focuses primarily on the insurgent group's use of violence and the population's threshold of violence. These concepts can be represented with two positively sloped lines (see Figure 1), with the lower line indicating the level of violence an insurgent movement must use over time as an indicator of its influence and ability to effect change, and the upper line indicating the maximum level of violence a population will tolerate before abandoning the group.

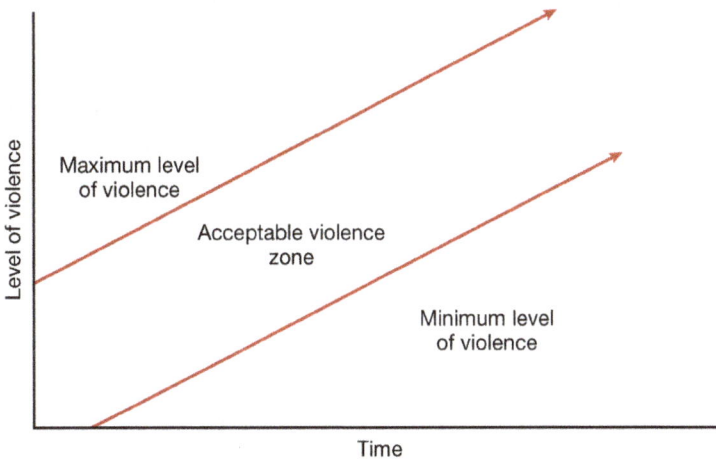

Figure 1. Threshold of violence.

However, more precision is required to fully understand the concepts. No host population will view an insurgent group's direct attacks as legitimate, so it makes little sense to talk about acceptable levels of violence waged by an insurgent group on a population. However, a populace may look favorably on violence committed by an insurgent group against what it sees as a hostile and illegitimate government. The downside, though, is that a certain level of insurgent violence may precipitate the government to wage a violent counterinsurgency campaign that inflicts significant costs (whether intended or not) on the population in question. Hence, for the lines indicating both the upper and the lower threshold in Figure 1, the violence in question is carried out by an insurgent group against a government the host population views as hostile or illegitimate.

The insurgent group faces a conundrum: It needs to use a minimum level of violence against its targeted government to project an image of strength and attract recruits. At the same time it cannot use too much violence because doing so could lead to an aggressive government response that could result in widespread devastation, which, in turn, may cause the population to become angry with the insurgent group for dragging it into a costly conflict. The canonical example is the reaction of the Palestinian populace in the West Bank and Gaza Strip to the patterns of violence initiated by the Israeli government and Hamas. Complicating matters is the difficulty of measuring the thresholds, particularly the upper threshold, because a community is not likely to spell out the exact amount of violence it is willing to tolerate. Rather, the upper threshold usually becomes clear through reaction or resistance, and one of the goals of an insurgent group may be to attempt to increase the upper limit over time.

This short paper begins with a brief review of the reasons insurgent groups engage in violence against a government; note that these reasons extend beyond simply gaining concessions from a government or eliminating it entirely. Then there is brief discussion of some of the forms of violence used by an insurgent group against a government, followed by some preliminary thoughts on strategies a movement may adopt if it wants to increase the upper threshold of the acceptable level of violence a population is willing to tolerate.

MOTIVATIONS FOR THE USE OF VIOLENCE

Clearly one of the motivations for an insurgent movement to use violence is the desire to force a government to make concessions, or indeed even a desire to eliminate a government entirely. A more subtle notion closely connected to this idea is that of "costly signaling." Essentially, in most cases an insurgency is much weaker than a government and therefore relies on terrorism and irregular techniques to signal to relevant audiences their resolve and ability to impose costs. This notion is captured by Kydd and Walter, who noted:

> Given the conflict of interests between terrorists and their targets, ordinary communication or "cheap talk" is insufficient to change minds or influence behavior. If al-Qaida had informed the United States on September 10, 2001, that it would kill 3,000 Americans unless the United States withdrew from Saudi Arabia,

the threat might have sparked concern, but it would not have had the same impact as the attacks that followed. Because it is hard for weak actors to make credible threats, terrorists are forced to display publicly just how far they are willing to go to obtain their desired results.[2]

Although this passage discusses costly signaling within the context of terrorist actions, the concept is also applicable to the irregular or unconventional techniques adopted by insurgent movements.

The literature on the use of violence by terrorist and insurgent movements also notes that there are other signaling functions associated with using violence. More specifically, signaling through attacks on highly symbolic targets can overcome what is known as the collective action problem. This problem is a fundamental one for most insurgencies, so it is explored at some length in this paper.

In the 1960s the economist Mancur Olson[3] noted that it is not necessarily the case that rational, self-interested individuals will act in support of a collective good or common interest even if doing so will better their individual situations. To take a concrete example, security is typically considered a collective or public good. In the context of international security, the United States has traditionally played an important role in providing security to Western Europe, particularly during the Cold War—so much so that it has often led to a concern among Americans that Europe was "free riding" on American defense spending by not spending a sufficient level of their own gross domestic product on defense.

The frustration American defense planners and politicians felt at low levels of European defense spending reflected frustration with the presence of a collective action problem with respect to the provision of the common defense of the Euro-Atlantic area. Olson noted a tendency toward the suboptimal provision of public goods, such as defense (especially as a group becomes larger) because individual members, believing that the collective good will be ensured anyway by the efforts of other members of the group, have an incentive to shirk their responsibility to provide for the public good. Such a strategy of free riding may be individually rational because often the consumption of a public good is available to all, not just to those who took action and paid the costs for its achievement. If many members of a group were to behave in this manner, this collective behavior would translate into the suboptimal provision of the collective good, and perhaps even its nonprovision.

How might the collective action problem help explain the mobilization challenges faced by insurgent movements? One can regard the desired outcome of an insurgency—such as a revolutionary socialist government, a territory under sharia law, a liberal democracy allied with the West, or an ethnocratic enclave carved out of a multinational empire—as a public good requiring self-sacrifice among the relevant population. Yet as McCormick and Giordano noted, insurgency is an inherently risky business that, in most cases, offers little probability of success.[4] More specifically, in the initial stages of a rebellion, an insurgent group tends to be small and less powerful than the state it is fighting. Under such conditions, the collective action problem is often insurmountable because most potential supporters (correctly) judge that the likelihood of success is remote, thus leaving only a small number of die-hard activists willing to risk their lives for a desired collective good. For this reason, most opposition groups die young.

How do groups that do go on to challenge the state for supremacy overcome this initial mobilization dilemma? One such strategy, as noted by McCormick and Giordano, is to wage violence against highly symbolic representations of state power. Such attacks, if successful, generate violent images that, whether factually correct or not, attest to the growing relative strength of the insurgents.[5] The process by which this happens, and the role of violence in effectuating this process, is captured by McCormick and Giordano (who quote Thomas Thornton):

> Violence . . . is used as an instrument of armed propaganda. The objective is to advertise the existence of an emerging opposition, raise popular consciousness and define the terms of the struggle. As Thomas Thornton has suggested, incumbents typically enter an insurgency in a natural state of political "inertia," even in the absence of significant popular support. The insurgents, for their part, begin the game as outsiders, an alien political force which "the organism of society will be predisposed to cast out." Before the opposition can even begin the process of building a base of popular support it must first be able to disrupt the system's inertial stability. "In order to do this, the insurgents must break the tie that binds the mass to the incumbents' by removing the structural supports that give [the system] its strength." These actions, as Thornton goes on to explain, will gradually sever the socio-psychological bonds that tie conditional elements of the population to the state and force them to choose between a disintegrating status quo and an emerging opposition. This cannot be achieved with words; it can only be achieved with violence.[6]

If highly symbolic acts are repeatedly carried out, the populace is then forced to choose between supporting what it perceives as a crumbling status quo and supporting an opposition that appears to be growing in strength. The natural choice is the latter, both as a result of a desire to support the winning side and partake in any spoils of victory and to avoid being on the receiving end of score-settling violence that may result from having supported the losing side (or not fully supporting the winning side).

Other scholars have noted other functions of violence against a state.[a] Jaeger et al. suggested that violence against Israel serves as a type of public good if indeed the Palestinian public desires to continue the violent struggle against Israel or if it simply has the desire for retribution.[9] Successful attacks by one Palestinian faction, such as Fatah or Hamas, could therefore lead to an increase in popularity for that group. Other scholars have noted that successful attacks reveal the "high quality" of a group, which signals that it will be successful in providing public goods.[10] Consistent with this idea, Berman and Laitin noted that radical groups are able to carry out effective campaigns of violence because the prohibitions they impose on members allow them to select only those most committed to the cause of the group, thereby weeding out those likely to shirk on their responsibilities.[11] Hence, as noted by Jaeger et al., "successful attacks against Israeli targets signal that the faction responsible has highly committed members and that those members will not be tempted by corruption and will deliver good governance in other dimensions of public activity as well."[12]

---

[a] Note that other scholars have advanced various explanations for why insurgent groups make threats or carry out attacks directly against populations. For instance, Kalyvas[7] suggested that insurgent violence against populations is a function of territorial control and information. Specifically, he argued that groups do not need to use violence against a populace if they already possess a high level of territorial control, and that lacking precise information on the identity and location of the enemy, they may wind up using indiscriminate violence. Wood noted that the Frente de Libertação de Moçambique attacked villages to demonstrate that Portuguese forces could not protect civilians. In general, Wood argued that groups with greater resources have the luxury of enticing populations with various forms of benefits and therefore do not need to use violence to compel loyalty or extract resources and recruits from a population. He stated, "In order to credibly and consistently deliver sufficient incentives to encourage and maintain large-scale civilian support, insurgents must possess some extant ability to control land, markets, or resources. Strong rebel organizations may be able to provide parallel political systems, public services, and similar incentives, but such goods exceed the capabilities of most rebel organizations. For this reason, weak insurgents are likely to be outbid by the government. Even in the context of high repression and low state capacity, incumbent regimes are likely able to offer a more competitive deal than insurgents. . . . Facing a highly unequal balance of capabilities, weak insurgent groups may view violence as an inexpensive alternative to supplying positive incentives to (temporarily) expand their resource base."[8]

TYPES OF VIOLENCE

The types of violence insurgent groups use include facility attacks, bombings, assassinations, kidnappings, hijackings, and violence against civilians. For example, during its nearly thirty-year campaign to politically separate the Tamil regions of the country from Sri Lanka, the Liberation Tigers of Tamil Eelam (LTTE) waged a terrorist and insurgency campaign that relied primarily on armed assaults and bombings, although a variety of other types of actions were undertaken (see Figure 2).

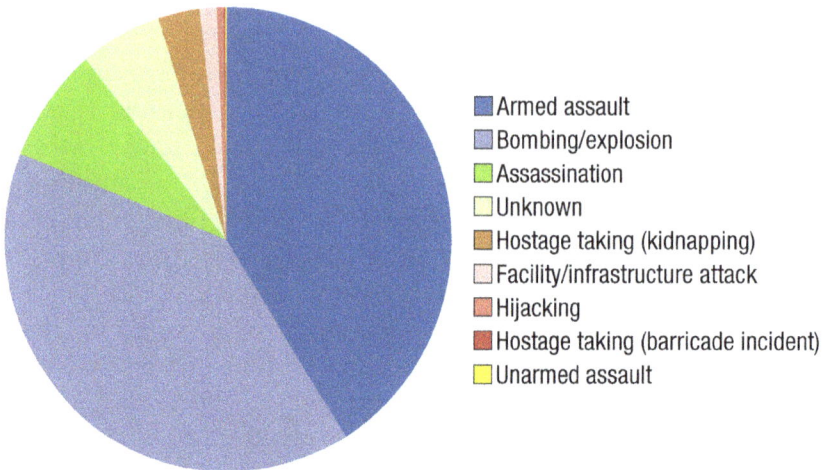

Figure 2. LTTE attack types.

Legend:
- Armed assault
- Bombing/explosion
- Assassination
- Unknown
- Hostage taking (kidnapping)
- Facility/infrastructure attack
- Hijacking
- Hostage taking (barricade incident)
- Unarmed assault

Additionally, the group employed a wide target list, as shown in Figure 3.

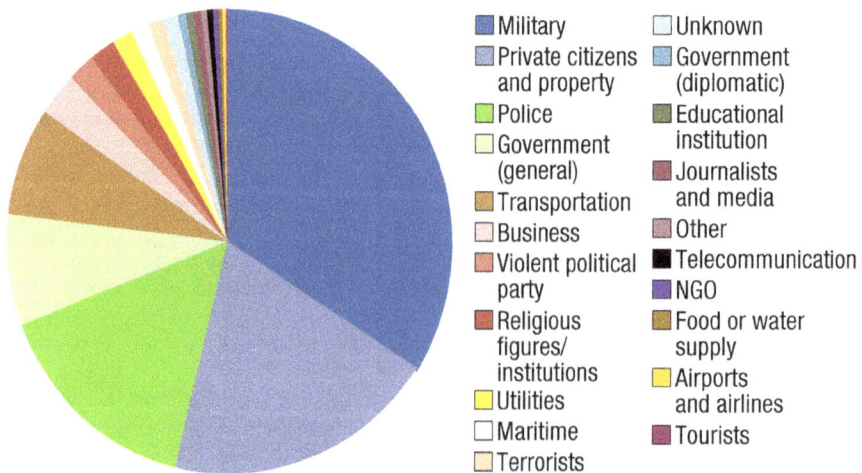

Legend:
- Military
- Private citizens and property
- Police
- Government (general)
- Transportation
- Business
- Violent political party
- Religious figures/institutions
- Utilities
- Maritime
- Terrorists
- Unknown
- Government (diplomatic)
- Educational institution
- Journalists and media
- Other
- Telecommunication
- NGO
- Food or water supply
- Airports and airlines
- Tourists

Figure 3. LTTE target types.

Hizbollah is also an interesting case. In an empirical study highlighting the group, researcher Luis de la Calle concludes "the repertoire of violence insurgents can use is determined by its capacity to seize and hold territory from the state's grip."[13] Groups without territorial control resort to terrorist tactics—such as improvised explosive devices, bank robberies, kidnappings, and assassinations—as they "maximize the impact of the act on the targeted audience." By nature of their small size and limited resources, weaker insurgent groups must stay underground, operate clandestinely, and avoid the full brunt of conventional military force from the incumbent; surprise "terrorist" attacks are thus the best course for their ends.

History reveals this trend among many insurgent groups in their nascent development stage, including Hizbollah, Algeria's National Liberation Front, Sri Lanka's LTTE, and South Africa's African National Congress. These groups waged campaigns of surprise attacks, public bombings, and urban terrorism. By contrast, stronger groups with territorial control are better equipped to employ guerrilla tactics—ambushes, raids, facility attacks, and small-scale battles—against the incumbent and have greater capacity to "build larger irregular armies and produce more deadly conflicts." The aforementioned groups, once they gained greater strength and more territory, changed their tactics from terrorism to guerrilla warfare.

Focusing on the case of Hizbollah, the insurgent group emerged in 1982 after the Israeli invasion of South Lebanon, and it transitioned its tactics from terrorism to guerrilla warfare as a direct result of its territorial gains and its increasing military capacity (with Iran playing an important role in supplying the group with military arms). Before 1991, the year that saw Hizbollah consolidate considerable political and military might, the group carried out the deadly 1983 suicide bombings of the American military barracks in Beirut, killing 241 American and 58 French troops.[14] The group also commonly kidnapped Westerners in the 1980s. Together, these two tactics represent "quintessential terrorist" actions for a rogue underground group battling materially stronger incumbents. The deputy general-secretary of Hizbollah, Naim Qassem, said: "Up until 1985, Hezbollah [*sic*] was not yet a single entity that could stand up and speak for itself . . . the nature of our formation required clandestine behavior."[15] After 1991, Hizbollah's tactics changed. With increased legitimacy among the Shiite community, more territorial control, and greater military capacity, the group started to "carry out raids and assaults against the Israeli Defense Forces that in some cases amounted to small scale battles that were sustained for hours."

When engaging an insurgent group militarily, it is therefore important to assess the group's level of territorial control and its resources in order to gauge its likely toolkit of violence. Groups with territory are more likely to attract more recruits (as a result of their access to the population), increase resources, and wage guerrilla-like warfare, including facility attacks, small-scale battles, and hit-and-run tactics. By contrast, groups without territory are forced to operate clandestinely and use terrorist tactics such as improvised explosive devices, assassinations, and kidnappings. As of the writing of this paper, the Islamic State of Iraq and the Levant (ISIL) is a forceful insurgent group gaining considerable swaths of territory in northern Iraq, a trend that has seen the group employ more guerrilla and conventional assaults. Conversely, the global Islamist militant group al-Qaeda operates a stateless army and wages large-scale terrorist attacks, such as the 9/11 attacks on the United States. Carried out by a network of followers, affiliates, and adherents, such acts are indicative of organizations without territorial control in the places they employ violence.

IMPACTING THE UPPER THRESHOLD
OF VIOLENCE

Following are some preliminary thoughts on how groups impact the upper threshold of violence. The threshold before the start of hostilities can be determined by a variety of contextual factors, such as the population's attitudes toward an insurgent movement and its causes and the degree to which the economic and other needs of the population are currently being met. It would behoove the leaders of a movement to pay attention to such contextual factors before engaging in hostilities with the enemy. A useful example is provided by the attitudes of the Shia population in southern Lebanon after the 1982 Israeli invasion in the southern part of the country to root out the presence of the Palestine Liberation Organization (PLO). It might seem logical that the Shia would side with their Muslim coreligionists in the PLO given the state of dispossession of the Palestinian people after Israel's creation in the late 1940s and another refugee crisis created after Israel's victory in the Six-Day War in 1967. However, it turns out that the Shia population was sick of the depredations of the PLO and had blamed the group for reprisal Israeli attacks before the 1982 invasion. In fact, the Shia in southern Lebanon initially welcomed the Israeli invasion, greeting incoming troops by throwing rice at them. Opinion would soon turn against the Israelis as it became apparent that they planned a long-term presence in the region. Nonetheless, the point is that the local population by the time of the 1982 invasion had a low (upper) threshold for Palestinian violence against Israel because such violence had resulted in reprisals that affected the Shia community. This low threshold, in turn, resulted from a revealed unwillingness of the population to bear the brunt of costs of PLO violence waged against Israel.

Perhaps one of the more important tactics a movement can rely on to raise the upper threshold is to goad a government into unleashing a counterinsurgent campaign that lacks restraint and, whether intentionally or not, leads to a significant number of deaths of noncombatants and destroys the overall economy of the host society. An obvious example is the radicalization of segments of the Palestinian population after wars with Israel that led to many noncombatant deaths and the devastation of the economy and infrastructure of the Gaza Strip. Bueno de Mesquita and Dickson noted that an indiscriminate counterinsurgent campaign can mobilize support for a group by signaling the government's unwillingness to offer concessions and by reducing the opportunity costs for participating in or simply supporting violent acts.[16] A reduction of opportunity costs refers to the notion that an indiscriminate counterinsurgency campaign is likely to lead to the widespread destruction of infrastructure and the local economy, and as livelihoods are destroyed, potential insurgents would not have to forgo earnings that they could have earned had there been a functioning economy during the conflict. An indiscriminate campaign may also address the free-rider problem, as those who are unsure who to support may conclude that they would have greater security by actually joining the insurgency and fighting the government.[b]

However, movements that adopt this strategy play a dangerous game. If the damage is extremely widespread, governing in the aftermath of the conflict may be impossible, as the widespread destruction of an economy and civil society are likely to hamper the ability to govern, and some segment of the population is likely to blame the group for dragging the entire society into an unwanted conflict. After hostilities in 2009, one despondent Palestinian woman whose house was destroyed stated:

> I will never vote for Hamas. They are not able to protect the people, and if they are going to bring this on us, why should they be in power? If I thought they could liberate Jerusalem, I would be patient. But instead they bring this on us.[18]

---

[b] As an accompaniment to the strategy of inducing an indiscriminate counterinsurgency campaign, a group may also suppress information regarding its role in bringing this situation about. For instance, during the July 2014 conflict with Israel, Hamas is reported to have threatened journalists for reporting that the organization used civilian sites to attack Israel. It has also urged social media users not to show evidence of rockets fired from population centers and to refer to all casualties as "innocent victims."[17]

More generally, Wood noted that the argument that a population traumatized by an indiscriminate counterinsurgency campaign would flock to the insurgents rests on an overstated assumption that the insurgents have the capacity to protect civilians and safeguard their livelihood, which may not be the case. He stated:

> While regime backlash may mobilize civilians who were already at or close to the point of indifference between remaining neutral and supporting the rebels, it does not necessarily follow that most civilians would support insurgents in the wake of indiscriminate regime violence. Without credible security guarantees from the rebels, civilians likely have insufficient incentive for supporting the risk of insurgents. Indeed, civilians may blame the rebels for the escalation in violence and withhold support. Even when government forces kill large numbers of civilians, destroy property, and use other forms of collective punishments, civilians choose to collaborate with the incumbent's forces if the rebels are seen as week.[19]

As an example, Wood cited the case of the Tigray People's Liberation Front (TPLF), an Ethiopian group that fought against a ruling military committee known as the Derg, which came to power after the overthrow of the Haile Selassie regime in 1974. Wood noted:

> The TPLF's experience in Afar is telling. The Derg's campaign of repression against the Afar generated significant grievances among the population but was insufficient to drive the people into the arms of the rebels. In order for the rebels to profit from Mengistu's violence, they had to credibly demonstrate their commitment to making positive contributions to the lives of the Afar. This meant providing benefits such as economic development, political and educational structures, security, and justice systems.[20,c]

The upper threshold of tolerance for violence can potentially be maintained at a high level if a movement is fortunate enough to have an outside sponsor willing to provide resources to pay for reconstruction at the end of hostilities. An example is Iranian sponsorship of Hizbollah. Iranian financing has played a vital role in rehabilitating

---

[c] Another example is the recent fighting between Israel and Hamas in Gaza. As one political scientist at Al Quds University in East Jerusalem put it, "All these achievements of Hamas, if they strike a deal without achieving something for the people of Gaza, they will lose everything and will bury themselves."[21]

areas of Lebanon devastated by Israeli attacks. For instance, Iran's Martyrs Foundation has partnered with a Hizbollah-affiliated organization known as Jihad al-Bina' (Holy Reconstruction Organ) to rebuild homes destroyed by fighting between Israel and Hizbollah. Representatives of the insurgent group determine the validity of families' housing needs and, if necessary, arrange for the required property transactions. Financing provided by the Martyrs Foundation is used to acquire land, with Jihad al-Bina' subsequently developing plans and building the finalized structure.[22] Reportedly, one month after the 1996 Israeli military campaign Grapes of Wrath, Jihad al-Bina' rehabilitated more than 2,800 structures damaged by Israel in 106 locations in South Lebanon,[23] and recent press reports indicate that Iran has spent $400 million to rebuild Dahiya, a southern suburb of Beirut, after the July 2006 Hizbollah war with Israel.[24]

The Martyrs Foundation has also made noticeable contributions to the provision of health care, particularly after hostilities. Specifically, it financed the construction of the al-Rasul al-Azam Hospital in Dahiya.[25] Injured Hizbollah fighters do not pay for medical expenses at this facility, while injured civilians pay 70 percent of their expenses. Such forms of outside sponsorship can potentially contribute to the maintenance of a high level of tolerance of a population for violence, particularly in the aftermath of hostilities.

There are other nonmaterial forms of rewards that a movement can offer to potentially maintain or increase the upper threshold of violence. Specifically, it could undertake a number of actions to raise the social status of the fallen, such as having its leaders make "Heroes Day" or "Martyrs Day" speeches commemorating the fallen, building monuments and painting murals dedicated to the fallen, or hanging oversized portraits of those who have given their lives for the movement. It may also decide to create a cult-like devotion to the notion of sacrifice or martyrdom, as was done by the LTTE and Iran's revolutionary leaders. The former argued that sacrifices and martyrdom were necessary to bring forward the day of victory, while the latter exalted the fallen by claiming they had a privileged place in paradise next to the Prophet Muhammad and cherished Shia imams, such as Imam Ali. Speaking of one recently fallen member of the Imam Ali garrison of the Iranian Revolutionary Guards Corps, one Shia theologian stated, "Where others were found by martyrdom, he looked for it. Praise be to God as he will have the fragrance of Imam Hussein."[26] Showing evidence of the resolve of the insurgent group is another potential tactic, as in the case of the famous cyanide pills all members of the LTTE wore in small capsules around their necks.

Lastly, many conflicts have a lengthy buildup, with governments and insurgents taking a variety of political and military actions in the belief that a conflict at some point in the future is likely. That being the case, to increase the upper threshold the leaders of an insurgency can, whether intentioned for this purpose or not, maneuver the group into a position where it is seen as moderate and willing to make political sacrifices while painting the opposing government as obdurate and itching for war. An *Economist* article on the recent fighting in Gaza noted:

> More recently, say Gazans, the Israelis under Binyamin Netanyahu showed they were determined to destroy a peace-minded Palestinian unity government endorsed by Hamas and the more moderate Fatah party under Mahmoud Abbas, after the failure of American-brokered talks between Mr. Abbas and Mr. Netanyahu. The Israeli prime minister made it clear that he would never talk to a Palestinian government backed by Hamas, even though America cautiously welcomed it. So he has done everything, say the Palestinians, to thwart it.[27]

Motivated by such perceptions, a population is likely to tolerate a high upper threshold of violence, at least in the initial stages of a conflict. As the *Economist* article also notes, "Firing rockets, many of them [residents of Gaza] argue, is the only way they can protest, even though they know the Israelis are bound, from time to time, to punish them."

NOTES

[1] Stathis Kalyvas, *The Logic of Violence in Civil War* (New York: Cambridge University Press, 2006), 19.

[2] Andrew H. Kydd and Barbara F. Walter, "The Strategies of Terrorism," *International Security* 31, no. 1 (2006): 50–51.

[3] Mancur Olson, *The Logic of Collective Action: Public Goods and the Theory of Groups* (Cambridge, MA: Harvard University Press, 2009), 1–2.

[4] Gordon H. McCormick and Frank Giordano, "Things Come Together: Symbolic Violence and Guerrilla Mobilisation," *Third World Quarterly* 28, no. 2 (2007): 295–320.

[5] Ibid., 309.

[6] Ibid., 307; and Thomas Thornton, "Terror as a Weapon of Political Agitation," in *Internal War*, ed. Harry Eckstein (Westport, CT: Greenwood Press, 1980), 74.

[7] Kalyvas, *The Logic of Violence in Civil War*, 12–13.

[8] Reed M. Wood, "Rebel Capability and Strategic Violence against Civilians," *Journal of Peace Research* 47, no. 5 (2010): 604.

[9] David A. Jaeger et al., "Can Militants Use Violence to Win Public Support? Evidence from the Second Intifada," *Journal of Conflict Resolution* 59, no. 3 (2015): 528–549.

[10] Ibid.

[11] Eli Berman and David D. Laitin, "Religion, Terrorism and Public Goods: Testing the Club Model," *Journal of Public Economics* 92, nos. 10–11 (2008): 1942–1967.

[12] Jaeger et al., "Can Militants Use Violence to Win Public Support?," 3.

[13] Luis de la Calle, "The Repertoire of Insurgent Violence," research paper, prepared for the American Political Science Association Conference, Seattle, September 2001.

[14] CNN Library, "Beirut Marine Barracks Bombing Fast Facts," *CNN*, June 13, 2013, www.cnn.com/2013/06/13/world/meast/beirut-marine-barracks-bombing-fast-facts.

[15] Hala Jaber, *Hezbollah: Born with a Vengeance* (New York: Columbia University Press, 1997), 62.

[16] Ethan Bueno de Mesquita and Eric Dickson, "The Propaganda of the Deed: Terrorism, Counterterrorism, and Mobilization," *American Journal of Political Science* 51, no. 2 (2007): 364–381.

[17] Ariel Ben Solomon, "Hamas Tells Social Media Activists to Always Call the Dead 'Innocent Civilians,'" *Jerusalem Post*, July 21, 2014; and Lahav Harkov, "Journalists Threatened by Hamas for Reporting Use of Human Shields," *Jerusalem Post*, July 31, 2014.

[18] Ethan Bronner, "Parsing Gains of Gaza War," *New York Times*, January 19, 2009.

[19] Reed M. Wood, "Rebel Capability and Strategic Violence against Civilians," *Journal of Peace Research* 47, no. 5 (2010): 604.

[20] Ibid.

[21] Jodi Rudoren and Ben Hubbard, "Despite Gains, Hamas Sees a Fight for its Existence and Presses Ahead," *New York Times*, July 27, 2014.

[22] Judith Harik, "Hizballah's Public and Social Services and Iran," in *Distant Relations: Iran and Lebanon in the Last 500 Years*, ed. H. E. Chehabi (London: I. B. Taurus & Co., 2006), 282.

[23] Ibid., 278.

[24] "Hezbollah Heartlands Recover with Iran's Help," *BBC News*, June 12, 2013, http://www.bbc.com/news/world-middle-east-22878198.

[25] Harik, "Hizballah's Public and Social Services and Iran," 275.

[26] Thomas Erdbrink, "With War at its Doorstep, Iran Sees Its Revolutionary Guards in a Kinder Light," *New York Times*, July 18, 2014.

[27] "Why Hamas Fires Those Rockets," *The Economist*, July 19, 2014.

INFORMATION TECHNOLOGY
IN INSURGENCY AND
RESISTANCE MOVEMENTS

Katharine Raley Burnett, James A. Gavrilis, and Theodore Plettner

Information technology is improving communications around the world. Today the Internet provides widespread, reliable, long-distance communications and technological access. Insurgents and resistance movements benefit from these improvements in communications technology and use existing popular networks, such as Facebook and Twitter, to support their activities.[1] The Internet, cell phones, and text messaging are all examples of reliable, long-distance, real-time communications systems that greatly enable insurgents and resistance movements today but were unavailable only a few decades ago. Insurgents and resistance movements use information technology to conduct operations on the battlefield, swell popular support, recruit new members, and finance activities.

These improved capabilities have strengthened insurgent groups and raise questions about the appropriate response of governments seeking to counter insurgent groups. Directly countering this technology, but limiting access to equipment and networks, is proving to be increasingly difficult. Much of the technology groups use is designed and distributed for the consumer market, making it nearly impossible to restrict access. Taking down websites and removing users from social media sites is legally challenging, and sites and accounts can be easily replaced. This primer will highlight some of the information technologies that insurgents and resistance movements use to enable their organizations and operations as well as the challenges in countering the use of these technologies.

Information technology, specifically social media, was used during the numerous revolutions throughout the Middle East between 2011 and 2015. Many journalists and scholars are examining the role of Facebook and Twitter in these revolutions and political upheavals, most notably during the Egyptian revolution ousting Hosni Mubarak in February 2011.

In a *New York Times* article, Jose Antonio Vargas describes "how the Egyptian revolution began on Facebook" when Wael Ghonim created the Facebook page "We Are All Khaled Said," using the name of a young man beaten and killed by Egyptian police.[2] His Facebook page epitomized the oppression felt by many in Egypt. In addition to providing a venue for expression, social media provided mass communications to the public. Rob Schroeder explains, "during the Egyptian protests, many users posted messages, images, and links to other web pages about what was occurring through Twitter. These helped create a global conversation about events in Egypt."[3]

Similarly, Tunisians were posting Facebook photos of events as they occurred, and they communicated using YouTube videos during the revolution in Tunisia in January 2011.[4] Closer examination is required to determine the exact impact of social media in relation to other factors that may have also played a role in the uprisings.

In the past, developed national militaries were the only organizations on the battlefield that had sophisticated communications capabilities such as short- and long-range audio, video, and data communications networks. These capabilities provided them with a marked advantage over the insurgents or resistance movements that used low-tech capabilities such as messengers and landline telephones. In contrast, today, insurgents and resistance movements use the same communications technologies as their formal military adversaries. Insurgents and resistance movements can use these communications systems, including smartphones and satellite phones, to pass information and coordinate operations and logistics.

The introduction of cell phones into Afghanistan created competition between the Taliban and the Afghan government as both sides attempted to use the new technology to their advantage. The Taliban has used text messages to spread propaganda and intimidate the population.[5] The Afghan people also use the cell phone network to anonymously inform the to Afghan government forces of the movements and activities of the Taliban. The Taliban has attempted to control the network by forcing providers to shut off phone service at night and attacking cell phone towers of companies that do not comply.[6]

Taliban informants also frequently used cell phone networks to improve the accuracy of indirect fire against US military bases and to notify insurgents of US patrol movements so insurgents could successfully place improvised explosive devices or ambush. They used these phone networks to provide early warning of an impending US military raid on an insurgent base.

In the 2008 Lashkar-e-Taiba attacks in Mumbai, the terrorists planned their attack using Google Maps and they carried GPS systems with them.[7] They were in constant cell phone contact with a command center that was monitoring the media coverage of the attack and providing the attackers with information about the identity of hostages. Whether individual hostages lived or died was determined by the results of a Google search.[8] Moreover, insurgents and resistance movements have used cutting-edge information technology successfully on the battlefield. For example, an Army press release stated:

> When a new fleet of helicopters arrived with an aviation unit at a base in Iraq, some soldiers took pictures on the flight line. From the photos that were uploaded to the internet, the enemy was able to extract the "geotags" and determine the exact location of the

helicopters inside the compound, and conduct a mor-
tar attack, destroying four of the AH-64 Apaches.[9]

The ability to communicate easily, quickly, and inexpensively with audiences near and far has greatly improved the ability of insurgents to recruit fighters, generate support among the population, gain necessary skills, and communicate and plan. The response of forces opposing these insurgencies must take these changes into account. Some governments have attempted to prevent access to information technology that facilitates rebellions, primarily by limiting access to the Internet or shutting down specific websites. The prospects for such actions in the future appear dim; increased access around the world means that adversaries will find new ways of communicating as soon as one is blocked.

There are also legal obstacles to stopping insurgent activity on the Internet. The US government attempted to prevent American groups from providing advice and training to the Kurdistan Workers' Party (PKK), in Turkey because the group is a Department of State-designated terrorist organization. The government eventually won its legal battle, but only after first losing one case in a federal court[10] before succeeding on appeal at the Supreme Court.[11] The process took six years, illustrating the limitations of relying on legal processes to restrict terrorists' and insurgents' use of the Internet.

Governments trying to counter insurgents and resistance movements within their borders may want to monitor or control cell phone communications in a given area, which can be difficult and costly. Cell phone networks are not established in many remote locations in Africa, the Middle East, or Central Asia, such as the Sahel, where many insurgents and resistance movements take root and grow. Nevertheless, new technologies are surfacing and becoming available to insurgents and resistance movements. In areas that do not have a large-scale communications infrastructure, new capabilities are emerging, such as Bluetooth bump technology and Wi-Fi mesh networks, that enable insurgents to communicate without cell phone networks. Governments are challenged in their abilities to trace and monitor these new communications technologies.

Information technology entrepreneurs are developing new ways to provide connections throughout the world to grow market space and support economic growth in other industries. For instance, the Kenyan company BRCK Inc. developed a battery-powered cellular router for use when power grids go out or are nonexistent. This router maintains connectivity in difficult conditions, transfers data, identifies good cell towers, can be repaired remotely, and is able to run a secure network in extremely austere locations.[12] Unfortunately, the same capabilities that support legitimate businesses and communities also enable insurgents and resistance movements. The al-Qaeda global jihad network is an example of this trend. Information technology enables al-Qaeda's network to pursue a transnational agenda across almost every continent. The organization's affiliated groups provide support to a number of local jihadist insurgencies in various countries across the globe.[13]

Information technology provides numerous venues in real time to allow disparate groups to conduct mutually supporting activities. For example, an insurgent leader can post future actions and intentions on Facebook for others to see and then reinforce, if they so choose. This hypothetical leader can also use an Internet bulletin board to communicate quickly and immediately worldwide. Furthermore, there are a host of software applications and techniques involving private encryption programs, coding photographs, and shared e-mail accounts that a user can leverage to hide messages and communications.[14] Insurgent and resistance movement intelligence operations have been greatly enabled by the new information technology tools available. These tools provide "cut outs" or hidden connections between insurgent groups and individuals, that challenge governments' ability to identify command and control relationships, if and when they exist. Information technology allows insurgent individuals and resistance movements to operate in a mutually supporting yet decentralized manner.

The Internet and other information technologies have enabled small insurgent groups and resistance movements to amplify their propaganda and draw in regional resources. With technological improvements in mass communications, insurgent groups and resistance movements can have a much greater and much faster effect than they could in the past because they can reach larger audiences more efficiently. The internet allows individuals and groups wishing to spread a message to reach large audiences quickly and inexpensively. In a conflict setting, the Internet has evened the playing field between insurgent groups and governments. Governments that previously could monopolize the television, radio, and newspapers are now challenged by groups capable of spreading text and video messages over the Internet to both the local population as well as a global audience.

In the years after the US invasion of Iraq, insurgents fighting the United States began using the Internet for propaganda purposes. Insurgent organizations e-mailed professionally produced newsletters and provided links to online videos demonstrating their ability to successfully attack US forces, as well as material justifying their tactics, such as suicide bombing, and glorifying martyrs. The combination of different materials produced a synergy that allowed the groups to recruit members and build public support.[15]

Because communication is faster, it can be argued that perceptions change faster and the tipping points of mobilization are reached more rapidly than in the past. The Syrian Electronic Army's fake tweet hoax is an example of the speed with which communication through new information technology can influence people's perceptions and actions. In April 2013, via a hacked news agency's Twitter account, the Syrian Electronic Army tweeted a fake news headline describing a bomb attack inside the White House. Panic spread quickly and Wall Street reportedly experienced $136 billion in losses.[16]

Some insurgents and resistance movements make cunning use of popular mass media, such as the international news networks, to carry their messages. This can potentially sway public opinion toward belief in the organization's mission and belief that the movement is greater (more influential) than it really is. In Iraq, US forces identified insurgents using websites for regional recruiting and propaganda.[17] In Afghanistan, North Atlantic Treaty Organization (NATO) forces identified the use of text messages to pass propaganda and intimidate the population.[18] The Taliban understands how important control of the population and the means of communications are to conflicts, and the group has warned that it may shut down civilian phone communications systems.[19]

Information technology such as Internet videos and DVDs allow for more colorful, lifelike, and comprehensive expression of propaganda themes. Video may be more impressive than a simple oral story, and it allows senior authorities to reach larger audiences than they otherwise could.[a] Anyone with access to the Internet can view propaganda videos online.[20] Individuals can view a video in private, removed from family or friends who may counter the propaganda. Moreover, information technology makes producing propaganda easier and faster than ever before. During the Russian annexation of Crimea, numerous videos were uploaded and viewed as part of an information campaign to support political and military actions there.[21] These videos complemented news broadcasts from television stations and undoubtedly shaped perceptions about the annexation.

A recent and very successful example of social media use for propaganda and recruitment is the Islamic State (IS). IS has produced rap videos, testimonials by soldiers, religious material, and videos of their successes in combat and grisly executions, all of which have been popularized through social media accounts. Many of the accounts promoting IS are run by individuals who have no real-world connections to the organization and who are connected only through the Internet. The quantity and diversity of online content has allowed for individuals

[a] Hamas's video of the graduation ceremony of its youth brigades and Fatah's video glorifying female suicide bombers are two examples.

93

to be radicalized entirely online, with people who have exhibited no history of Islamic extremism surprising family and friends by suddenly traveling to Syria to join IS.[22]

Information technology can help people build and expand networks. Insurgent groups and resistance movements may be empowered to stay connected to external support networks across borders and throughout their region. Although diasporas have been a resource for insurgents for many years,[23] today's information technology enables insurgents to garner popular support from diasporas from around the world at much greater volumes, speeds, and frequencies than in the past.

Insurgents are sometimes able to raise funds by using these networks. Information technology has played a role in increasing insurgent financing. Information technology has enabled insurgents and resistance movements to embed in the myriad of black and gray markets and illicit financial networks, turning insurgency into a profitable endeavor.[24] For example, in Iraq in 2005 insurgents generated large amounts of revenue through gas smuggling, money laundering, narco-trafficking, kidnapping, and extortion, creating a "shadow economy" that undermined development of Iraq's legitimate markets as well as the government itself.[25] The Iraqi insurgency has become very profitable, and donors from Syria have received a large return on their investment.[26] Jihadists in Nigeria and northern Mali are now smuggling cocaine to Europe, which has generated millions of dollars more than their kidnapping activities garnered for their organizations.[27]

In addition to supporting the financial activities of insurgents or resistance movements, information technology supports other forms of external support. Political, psychological, and moral support can travel through social media with greater regularity than before. Regardless of where insurgents are, they now has vast amounts of knowledge available to them the Internet, including information about how to make improvised explosive devices, how to make detonation devices, how to resist detention, and how to conduct guerrilla tactics and strategies. Individual insurgents are greatly empowered by this access and connectivity.

There are drawbacks to using information technology for insurgent activities. For instance, if the government is able to identify the insurgent or revolutionary, it can take swift action. This was the case in Iran during and after the Green Revolution. The Iranian government blocked social media networks to thwart protests, used text messages and e-mails as evidence against protesters and organizers, and monitored the calls of Iranians participating in the Green Revolution. These government techniques appear to have been successful because reports of detention, imprisonment, torture, and murder of protesters and opposition leaders followed.[28]

The difficulty governments face when responding to insurgents' use of the Internet is highlighted by the ongoing debate and legal battle involving humanitarian assistance groups in California that were providing advice and training to the PKK in Turkey, a US Department of State–designated terrorist organization. A federal judge in California ruled that it was unconstitutional to ban "giving expert advice and assistance to terrorist organizations."[29] Later in 2010, the US Supreme Court ruled that it is illegal to provide any kind of assistance to terrorist organizations.[30] This ruling sparked a great deal of backlash from groups who say that it prevents humanitarian organizations from helping to find peaceful settlements between warring factions.

The legality of government responses to insurgents' use of information technology requires much more examination if governments are going to be capable of dealing with such use. It appears that it may not be effective for a government to try to counter or block any one form of insurgent communication, because the effects are short-lived and the insurgent or resistance movement can substitute other forms of communication. Governments will need more sophisticated techniques to identify the user, insurgent, or movement and then take action against the individual or group and not the form of information technology used. In addition, governments will need to come together to develop laws and enforcement mechanisms, both national and international, to instill more accountability and discipline in the use of national, regional, and international communications systems.

Information technology is continuing to advance, and insurgent groups will undoubtedly find ways to adapt new technologies to their struggles. To respond to the increased capabilities of these groups, the United States and other nations interested in countering insurgent activities will have to develop the ability to exploit insurgents' use of technology, adapt to the changing information environment of the Internet, and increasingly engage with worldwide populations targeted by online propaganda. More research is needed to explore specifically how information technologies are enabling these movements. In particular, it would be useful to explore how information technologies are improving the functionality of insurgent organizations and impacting popular sentiments about resistance movement objectives and the overall influence of these technologies on the outcomes of conflicts and revolutions.

NOTES

[1] Shaun Waterman, "Terrorists Discover Uses for Twitter," *Washington Times*, April 28, 2011.

[2] Jose Antonio Vargas, "Spring Awakening: How an Egyptian Revolution Began on Facebook," *New York Times*, February 17, 2012.

3 Rob Schroeder, Sean Everton, and Russell Shepherd, "Mining Twitter Data from the Arab Spring," *CTX2* no. 4 (November 2012).

4 Ethan Zuckerman, "The First Twitter Revolution?" *Foreign Policy*, January 14, 2011.

5 "ISAF Discusses Insurgent Propaganda Messaging," International Security Assistance Force – Afghanistan Headquarters press release, August 5, 2010, http://www.isaf.nato.int/article/isaf-releases/isaf-discusses-insurgent-propaganda-messaging.html.

6 Jon Boone, "Taliban Target Mobile Phone Masts to Prevent Tip-offs from Afghan Civilians," *The Guardian*, November 11, 2011.

7 Gabriel Weinmann, "New Terrorism and New Media," The Wilson Center Research Series, vol. 2 (Washington, DC: Commons Lab of the Woodrow Wilson International Center for Scholars, 2014).

8 Marc Goodman, "How Technology Makes Us Vulnerable," *CNN*, July 29, 2012.

9 Cheryl Rodewig, "Geotagging Poses Security Risks," US Army press release, March 7, 2012.

10 David Rosenzweig, "Judge Rules Against Patriot Act," *Los Angeles Times*, January 27, 2004.

11 Chris McGreal, "U.S. Supreme Court: Nonviolent Aid to Banned Groups Tantamount to 'Terrorism,'" *The Guardian*, June 21, 2010.

12 Heidi Vogt, "Made in Africa: a Gadget Startup," *Wall Street Journal*, July 10, 2014.

13 Seth Jones, "Resurgence of al Qaeda," *RAND Review* 36, no. 2 (Fall 2012).

14 Weinmann, "New Terrorism and New Media."

15 Jake Tapper, "Iraqi Insurgents Increasingly Use Internet as Propaganda Machine," *ABC News*, February 14, 2006.

16 Weinmann, "New Terrorism and New Media," 8.

17 Tapper, "Iraqi Insurgents Increasingly Use Internet as Propaganda Machine."

18 "ISAF Discusses Insurgent Propaganda Messaging."

19 Jon Boone, "Taliban Target Mobile Phone Masts to Prevent Tip-offs from Afghan Civilians," *The Guardian*, November 11, 2011.

20 Laura Collins, "'Jihad Cool': The Young Americans Lured to Fight for ISIS Militants with Rap Videos, Adventurism, and Firsthand Account of the 'Fun' of Guerrilla War," *Daily Mail Online*, June 19, 2014.

21 "Putin Awards Russian TV for Crimea Coverage: Kremlin-Controlled Media Accused of Propaganda Role," *Ukraine News One*, May 6, 2014.

22 Collins, "'Jihad Cool.'"

23 Daniel Byman et al., *Trends in Outside Support for Insurgent Movements* (Santa Monica, CA: RAND, 2001).

24 David Grange and J. T. Patten, "Assessing and Targeting Illicit Funding in Conflict Ecosystems: Irregular Warfare Correlations," *Small Wars Journal* (September 2009).

25 Robert E. Looney, "The Business of Insurgency: the Expansion of Iraq's Shadow Economy," *The National Interest* (Fall 2005): 70.

26 Bilal Wahab, "How Iraqi Oil Smuggling Greases Violence," *The Middle East Quarterly* 13, no. 4 (Fall 2006): 53–59.

27 Rachel Ehrenfeld, "Drug Trafficking, Kidnapping Fund Al Qaeda," *CNN*, May 4, 2011.

28 "Iran: Halt the Crackdown: End Violent Attacks on Protesters, Arrests of Critics," *Human Rights Watch*, June 19, 2009. Also see Narges Bajoghli, "Iranian Cyber-Struggles," *The Middle East Research and Information Project*, May 3, 2012.

29 Rosenzweig, "Judge Rules Against Patriot Act."

30 Chris McGreal, "U.S. Supreme Court: Nonviolent Aid to Banned Groups Tantamount to 'Terrorism,'" *The Guardian*, June 21, 2010.

THREAT FINANCE

Timothy Wittig

Portions of this work were adapted from Chapter 4, "Financing," in *Undergrounds in Insurgent, Revolutionary, and Resistance Warfare*, 2nd ed., ed. Robert Leonard (Fort Bragg, NC: United States Army Special Operations Command, 2012), 57–73.

INTRODUCTION

From the Taliban in Southwest Asia to al-Shabaab in the Horn of Africa to drug-trafficking organizations in Mexico, the behavior, capabilities, and ultimate success or failure of terrorist, criminal, and other transnational threats are closely tied to economic and financial factors. This close link among an adversary's operational, strategic, and financial dynamics is widely recognized and has spawned a boutique art aimed at countering and taking advantage of what has come to be known as "threat finance" within not only the Special Operations Forces (SOF) but also within conventional military units, interagency counterterrorism and intelligence efforts, law enforcement, and the financial and charity sectors.

Finances are the lifeblood of an insurgent or revolutionary movement. Wise insurgent leaders think of financial matters as a key and defining activity of the overall strategy and ideology of the movement, rather than as a peripheral matter. The nature of the movement's financial strategies tends to color the entire organization, in some cases in ways that undermine its founding ideology. When an insurgency dips into the lucrative illegal drug trade, it attracts attention. When an insurgency robs or extorts citizens within a country, it gains a reputation as a criminal organization. When a group subsists off of substantial funding from a foreign power, it tends to be viewed as a puppet organization taking its cues from that distant power. Likewise, the manner in which a movement handles acquired funds tends to characterize the organization in the eyes of the indigenous population. Insurgencies that distribute funds to impoverished citizens gain favor as the champions of the underprivileged. Conversely, leaders who succumb to the temptation of corruption tend to bring discredit on their organizations.

An economically sustainable insurgency is complex and must balance a number of critical factors. The insurgency must raise sufficient funds to sustain its operations and activities, but it must do so in ways that do not undermine its popular support. Undergrounds must do business with a myriad of people and suppliers, many of which are not sympathetic or are even antagonistic to the group's cause. Terrorists, insurgents, and revolutionaries today must not only be entrepreneurs and businesspeople, but they must also know how to navigate and move smoothly between informal traditional local economies and highly formalized Western commercial systems. In the places where most insurgencies operate today, informal business is the norm, and trade and business are based on trust and long-term relationships—often within a particular cultural group. In these places, laws and regulations tend to be flexible, governments corrupt, patronage rife, and paperwork easily made to say whatever it needs to. Learning to understand and leverage for advantage how militant underground movements navigate these dynamics is critical to determining successful SOF strategies.

When cultivating and supporting an underground, developing sustainable financial strategies and tactics can often be just as important as any other activity. Conversely, government forces conducting counterinsurgency must examine sources and management of insurgent finances as part of their overall strategy to defeat the movement. On a strategic level, exploiting an insurgency's financial vulnerabilities can negatively impact an adversary's operational, social, political, economic, or other capabilities and can provide friendly forces with critical insights regarding the adversary's behaviors, relationships, trends, and strategies. Tactically, exploiting threat finance dynamics can simultaneously help identify an insurgent group's critical capabilities and uncover its (sometimes hidden or counterintuitive) vulnerabilities. In particular, information about a threat's finances can be leveraged to

- disrupt enemy supply lines for needed equipment, people, or services;
- shape the adversary's behavior to create tactical windows of opportunity;
- degrade the adversary's specific operational capabilities;
- disrupt the enemy's exploitation of local economy;
- reduce the enemy's disruption of development projects;
- strengthen the local population's independence from enemy-connected value chains;
- undermine the enemy's relationships with patrons, proxies, and the population;
- discredit the enemy's objectives, operations, and propaganda; and
- create or validate advantageous propaganda narratives, among other things.

Threat finance is a critical issue for both undergrounds and SOF. Learning to understand economic and financial dynamics and leverage them for effect is a critical and perhaps underused weapon in our national arsenal.

FUNDING MECHANISMS OF RESISTANCE GROUPS

Nothing can happen for insurgent movements without money. Financing revolutionary movements is a key underground function, and the very methods of obtaining funds have a direct impact on the nature, ideology, and strategy of the movement. It is conceivable that a spontaneous resistance movement might emerge from public dissatisfaction, a protest march, or a riot, but sustaining and growing a movement into something that will effect change requires time, patience, and, above all, money.

Depending on their activities, undergrounds may need money to pay salaries, to finance advances in support of missions, and to purchase communication equipment and materials (e.g., Internet access and propaganda publications) and materials for sabotage (e.g., explosives), among other things. An underground may also extend aid to families who shelter refugees to help defray the families' costs; this occurred in Belgium during World War II after the Nazis eliminated many resistance collaborators from the bureaucracy. Previously, these sympathizers supplied fugitives with documents enabling them to switch identities and hold jobs. When these collaborators were removed from their positions, many evaders were forced into hiding, and money for their care was supplied by the treasury of the Armée de Belgique.

Financial aid may also be extended to the families of underground workers who have been captured or forced to flee. An example is the support the Luxembourg resistance provided to the dependents of 4,200 persons who were deported and nearly 4,000 who were sent to prisons and concentration camps during the Nazi occupation. L'Oeuvre Nationale de Secours Grande-Duchesse Charlotte not only provided immediate care for orphans but also gave each a thirty-thousand-franc trust fund. At the same time in Belgium, the Mouvement National Belge began the Fonds de Soutien (Funds for Support) for the families of workers in hiding. Similarly, Hamas (or Harakat al-Muqawama al-Islamiyya) and other Palestinian resistance movements offer families of suicide bombers or other "martyrs" financial compensation as a means of encouraging other impoverished families to produce fighters for the cause.

Money is also needed for bribery. Insurgencies thrive under corrupt governments, and undergrounds often disperse money to key officials to obtain their protection or silence. Bribery also plays a part in subversion and intelligence gathering. In most areas of the world, money equates to power and influence.

An underground may also channel funds to military units to pay salaries and buy supplies. In the Philippines, one of the prime responsibilities of the Communist Politburo in Manila was to obtain money for the Hukbalahap movement; and in Malaya, the Min Yuen was the major supplier of money to the rebels, often obtaining funds by extorting money from large landowners and transportation companies and appropriating cash from Communist-dominated unions. Hizbollah in Lebanon required money to maintain a sizable guerrilla force, along with missiles aimed at Israel. In a similar manner, undergrounds that support a guerrilla component require reliable funding sources.

Some insurgencies require funds to support their social outreach work and shadow government activities. Just as legitimate state governments struggle with the rising cost of medical care, unemployment insurance, food aid, housing subsidies, and pensions, some insurgent movements also struggle to provide similar services in an attempt to undermine the government, care for their constituents, and provide a cover for illegal and violent activities. These activities are expensive and require sustained and reliable income.

The cost of high-profile individual acts of violence tends to be miniscule compared with the overhead costs of simply sustaining and administering the underground, auxiliary, armed component, and public component. Spectacular acts of terror can inspire increased devotion and donations from supporters for a relatively low cost. But the day-to-day administration required to sustain and grow a movement is considerable.[1] Undergrounds obtain financing through a combination of external and internal sources.

EXTERNAL SOURCES OF MONEY

Foreign Governments

Often an underground is aided by an outside sponsor, usually a government, but more recent examples include sponsorship by non-state actors. Much of the money used by the anti-Nazi Belgian resistance of World War II, for example, came from franc reserves in London released by the British government. At one time, these reserves provided ten million francs per month. Similarly, many of the funds used by the French resistance were remitted from the Bank of England or sent from the Bank of Algiers (after the Allied landing in North Africa). The Viet Cong's resistance against the government of South Vietnam and its American allies was funded by both China and the Soviet Union, as well as by the authorities in Hanoi. Some external sponsors, such as Colonel Muammar Qadhafi in Libya, supported a number of different insurgencies. Qadhafi provided funding to groups ranging from Charles Taylor and the Revolutionary United Front (RUF) in Liberia and Sierra Leone, respectively, to the Provisional Irish Republican Army (PIRA) in Northern Ireland. However, other sponsors, such as Iran, tend to focus their sponsorship funding on insurgencies that advance the interests of Shia populations and those that undermine Israeli (and Western) influence, such as the Palestinian cause.

Foreign governments extend support to undergrounds for several reasons. The most important reason is that the activities of an underground often contribute to the defeat of a common enemy. Such aid also enables the sponsor to demand some reciprocity on the part of the underground. However, an outside government may give financial assistance to an underground even if there is no common enemy. According to one report, such a case occurred in 1940 when the Japanese government—not yet allied formally with Germany and Italy—provided the Polish underground in Rome with financial aid as well as technical equipment and Japanese passports in exchange for intelligence data on the German and Soviet occupying forces.[2]

Non-State Actors

In addition to governments, friendship societies or quasi-official aid groups may channel funds to an underground. Perhaps the best known of the latter was the Jewish Agency, which, during the Palestine revolution, had offices or representatives in every part of the Western world. In the run-up to the Israeli War of Independence, Palestinian Jews obtained critically needed financing from fellow Jews throughout the world, especially in Europe and the United States. Open appeals for money were made in newspapers and lectures and at charity balls and other social events.[3]

The vast majority of the Liberation Tigers of Tamil Eelam (LTTE) financing came from the large Tamil expatriate community, especially those contingents in Western countries (Canada, the United Kingdom, Australia, the United States, and Scandinavia) but also those living in the Indian province of Tamil Nadu. Indeed, many analysts identify the overseas Tamil communities as the single most important actor enabling the insurgency. Expatriate support included voluntary contributions from individuals and Tamil-owned businesses, as well as funds extorted from expatriates. LTTE collection methods evolved over time, from poorly coordinated, often violent acts of coercion to scheduled collections managed by computerized databases that allowed overseas collectors to avoid paying visits to individuals who supported rival Tamil groups or who were already regular contributors. Collections were made monthly or annually; additional collections were made according to special dates commemorating specific battles or individual "martyrs." Information was also collected on extended families residing elsewhere in order to lend credibility to threats against those who did not make expected donations. A 2009 Canadian intelligence report revealed that the community of expatriates and sympathizers within Canada was one of the top contributors to the LTTE, with donations of approximately

$12 million per year. After the LTTE lost control of the Jaffna peninsula in the mid-1990s, this source of financing became increasingly important, by some accounts providing up to 90 percent of the group's operating funds. However, the classification of the LTTE as a terrorist group by most Western countries, the result of an intense lobbying effort by anti-LTTE forces, put a serious strain on LTTE's ability to raise and transfer expatriate funds and was probably a major contributor to the Tigers' defeat in 2009.[4]

The main source of overseas funding for the New People's Army (NPA) in the Philippines during the Ferdinand Marcos administration was humanitarian organizations, including a number of European churches, and radical groups in Europe. The Communists, working through their public face, the National Democratic Front (NDF), touted the NDF as the only viable opposition to the human rights abuses of the Marcos regime. By 1987, the NDF had attracted the support of numerous international human rights organizations and established support networks in more than twenty-five countries. The NPA strategy was to divert resources from the nongovernmental organizations (NGOs) through aboveground institutions that were run by NPA supporters under the auspices of rural aid and development. These organizations remained a major source of support even after the fall of Marcos.[5]

PIRA likewise relied on support from abroad. The group turned to the United States for money and weapons as soon as it was organized enough to send agents abroad, and the Irish communities of Boston and New York proved very supportive. In 1969, the Irish diaspora in the United States was five times the size of the population of Ireland. The Irish Northern Aid Committee (NORAID) was set up in New York City in 1970 to provide a steady stream of money to the IRA, mostly for the purchase of weapons.[6]

Charities and nonprofit organizations are attractive sources of funding for insurgent undergrounds because they tend to be less regulated and scrutinized than publicly owned corporations, and some have a global reach and presence, administering considerable sums of money. Insurgencies can abuse charities through fraud (e.g., appropriating monies from a real charity intended for charitable purposes and using them to fund the movement instead); through a sham organization (i.e., creating an entirely bogus charity that poses as legitimate); or through co-opting a charity's money (i.e., using the money for the purpose intended by the donors but administering and distributing those funds through the insurgency).[7]

Cash in the Local Currency

Aid is often given in the form of cash in the local currency, which has the advantage of being easily exchanged for goods or services. The main problem is the physical transfer of the money. Usually this is handled by a front business organization through diplomatic channels, clandestine couriers, or infiltrated agents. A growing concern and challenge for counterinsurgency organizations, however, is the expanding ability to transfer money internationally using informal funds transfer (IFT) systems, such as the *hawala* system in the Middle East, *fei-ch'ien* in China, *hundi* in India, and *padala* in the Philippines. These systems can provide cash in the local currency for the recipient and are discussed in more detail in the "Parallel Financial Systems" section.

Substitute Currency

Stable, widely convertible ("hard") currency, such as US dollars or British pounds, is sometimes given to an underground when the sponsoring government lacks adequate reserves of the local currency. Hard currency makes a good substitute because it is easily exchanged on the black market for local currency or goods. Hard currency is also useful when the local currency is confiscated by the authorities and replaced by scrip, a frequent government countermeasure. The Castro regime took this measure soon after the Cuban Revolution.

Dollars were used extensively in financing World War II undergrounds. One British agent in Yugoslavia reported that it was no trouble to use dollars (or gold pieces) because "there was invariably a market for 'good' money in the towns."[8] More recent developments include the potential use of electronic credits accrued and traded via mobile phones as a means of transferring funds from one location to the next, especially in East Africa where the use of this type of electronic currency is commonplace and the cash value for online video-gaming credits can be acquired and traded with little visibility.[9]

Counterfeit Money

One other way to finance an underground movement is through the use of counterfeit money. For example, Chechen separatists established multiple production facilities in cities and mountainous regions to produce counterfeit US dollars. Although production of counterfeit money is not exclusively the province of a sponsoring government, undergrounds usually lack the necessary facilities and technical competence to produce counterfeit money; therefore, the main effort is generally undertaken by friendly governments.

Of course, the use of counterfeit money adds to the dangers underground members already face. During World War II, this factor reportedly prompted the Polish state underground to reject an offer of counterfeit money from London. It is possible that advances in digital printing technology could encourage insurgent groups to turn more to counterfeiting as a revenue source in the future.

Online Fraud

Insurgencies increasingly use illegal online operations to steal money or goods. Techniques include credit card and online banking fraud. In some cases, insurgents purchase stolen credit card or bank account numbers and passwords from criminal organizations and then use that information to withdraw money from compromised accounts or to buy goods directly. This avenue of obtaining funds illegally highlights an ongoing conflict between insurgents exploiting vulnerabilities in global computer networks and various government and international organizations attempting to fix those vulnerabilities and shut down online fraud through technical means, legislation, and enforcement.

Parallel Financial Systems

Islamic history and culture gave rise to an innovative and effective approach to financing insurgency as practiced by members of the Muslim Brotherhood and al-Qaeda. Hasan al-Banna, the founder of the Muslim Brotherhood, viewed finance as a critical weapon in undermining the infidels and reestablishing the Islamic caliphate. To meet these goals, he believed Muslims needed to create an independent Islamic financial system that would parallel and later supersede the Western economy.[10] Al-Banna's successors set his theories and practices into motion, developing uniquely Islamic terminology and mechanisms to advance the Brotherhood's system of faith, as well as its unique financial apparatuses.

President Gamal Abdel Nasser negated the Brotherhood's attempt to establish an Islamic banking system during the mass arrests in 1964. But Saudi Arabia welcomed this Egyptian dissident idea, and in 1961, King Saud bin Abdul Aziz funded the Brotherhood's establishment of the Islamic University in Medina to proselytize their fundamentalist Islamic ideology. In 1962, the Brotherhood convinced the king to launch a global financial joint venture that established numerous charitable foundations around the world. This joint venture became the cornerstone of the Brotherhood and was used to spread Islam (and later to fund terrorist operations) worldwide. The first of these charitable organizations were the Muslim World League and Rabitta al-Alam al-Islami, which united Islamic radicals from more than twenty nations. In 1978, the kingdom backed another Brotherhood initiative, the International Islamic Relief Organization, an entity that has been implicated in funding organizations such as al-Qaeda and Hamas.

Most Muslim nations collect mandatory Islamic charity (*zakat*) of approximately 2.5 percent of annual income from Muslim institutions and companies.[11] *Zakat* is intended to go to those who are less fortunate. However, the Brotherhood determined that those engaged in jihad against the enemies of Islam are entitled to benefit from the charitable offering. Muslims widely accept the interpretation that modern jihad is a serious, purposefully organized work intended to rebuild Islamic society and state and to implement the Islamic way of life in the political, cultural, and economic domains, and thus those involved in jihad are viewed as legitimate recipients of *zakat*.[12]

The rise of global militant Islam, and al-Qaeda in particular, also benefited from the use of parallel financial systems. The roots of al-Qaeda lie in a number of different organizations, including the Maktab Al Khidamat (MAK) or Services Office, a clearinghouse established to facilitate the recruitment, transportation, organization, training, and equipping of Arabs to support the Afghan resistance. Established by Abdullah Yusuf Azzam, a Palestinian scholar of Islamic law, and Osama bin Laden in Peshawar, Pakistan, in 1984, the MAK also attracted other militant leaders.[a] The MAK consisted of a network of international recruiting offices, bank accounts, and safe houses and was also responsible for the construction of paramilitary camps for the training of militants and the fortifications used by Arab fighters.[13,b] Between 1982 and 1992, estimates report approximately thirty-five thousand foreign fighters contributed to the Afghan effort, although there were probably never more than two thousand in Afghanistan at any one time. The

---

[a] Azzam, who earned a PhD from Cairo's Al-Azhar University, was a charismatic character and played a central role in crafting the narrative of resistance that drew thousands of Arabs to the Afghan cause. His previous combat experience (he fought the Israelis in 1967), combined with his religious credentials (his education and his connections with the family of Sayyid Qutb, an important ideological leader of the early Muslim Brotherhood), made him a particularly appealing figure to bin Laden. Omar Abd Al-Rahman, an Egyptian militant also educated at Cairo's Al-Azhar University and a key ideological figure for both Al-Jama'at Al-Islamiyya and the EIJ, also used MAK's resources to contribute to the Afghan resistance.

[b] Azzam originally established the MAK and later persuaded bin Laden to join. Bin Laden used his family's relationship with the Saudi royal family to support the effort overtly—through a strategic communications plan—and covertly, eventually matching US financial contributions to the resistance.

MAK was responsible for training approximately twelve thousand to fifteen thousand of those fighters, with approximately four thousand remaining connected through either chain of command or ideological affinity after the conflict.[14] To complicate the tracking of finances, leaders used parallel financial systems drawn from historical Islamic roots.

Both the Egyptian Islamic Group (EIG) and the Egyptian Islamic Jihad (EIJ) likely used these mechanisms to their advantage. EIG often sought donations during Friday prayers at mosques to facilitate its social outreach programs. It is unknown how much of this funding went toward illicit activity; however, it is presumed that the preponderance of the collected funds was reinvested into public and legal organizational activities. The extent of the aid EIJ received from outside of Egypt is not known, although the Egyptian government has claimed that both Iran and Saudi Arabia have provided financial and material support to EIJ.[15]

Given the prevalence of obtaining funds through such means, it may be assumed that EIJ obtained some funding through various Islamic NGOs, cover businesses, *zakat* funding operations, and possibly, although not likely, criminal acts.[16] The most evident external support to EIJ was its symbiosis with al-Qaeda and its increasing dependence on that organization. Few exact figures exist; however, from 1996 to 1997, EIJ received more than $5,000 per month[c] from al-Qaeda.

In the *hawala* system, money is transferred from a worker in one country to a worker in another country through the use of intermediary *hawaladars* in each country. Although a fee (cash, goods, or services) is charged for this transaction, the fee is usually much less than that charged in the formal banking sector, and this system also allows for the transfer of funds to and from countries and regions with limited financial infrastructure.[17]

INTERNAL SOURCES OF MONEY

Noncoercive Means

Gifts

Voluntary gifts from wealthy individuals and, occasionally, from commercial enterprises have constituted a good source of income for

---

[c] In late 1996, Dr. Ayman Al-Zawahiri traveled clandestinely to a number of former Soviet Caucasian republics (including Chechnya) and in December of that year was arrested by a Russian patrol in Dagestan. He was released in May, having stuck to his cover story without being identified by the Russians; however, he was chastised by al-Qaeda members for his carelessness and saw the subsidy (paid to EIJ by al-Qaeda) lowered to $5,000 during his imprisonment.

many undergrounds and are easier to hide from security forces. A few wealthy Chinese businessmen in Manila made large gifts to the Hukbalahap; the Malayan Min Yuen received substantial aid from several Chinese millionaires in Singapore. Many industrialists and bankers provided funds for the anti-Fascist underground in Italy. However, donor firms in France during the resistance encountered difficulties in hiding their donations from the Germans, and this hampered the exploitation of this potential source of revenue. In addition, financial gifts to the underground can also come from friends and relatives of underground workers and, given the manpower and opportunity, an underground may canvas door to door for contributions. Finally, dues levied on underground members can also provide needed funds.

Loans

The underground may also borrow funds. The Yugoslav Partisans, for example, floated a twenty-million-lira loan, which was marketed among the Slovene populace as "Liberty Loans"; and Belgian banker Raymond Scheyven's Service Socrates organization, in the name of the government-in-exile, managed to borrow more than two hundred million francs for the anti-Nazi Belgian underground from the end of 1943 until the Allied liberation of Belgium.

An underground worker attempting to solicit funds from strangers is sometimes challenged to convince them of the agent's good faith. The underground worker may provide strangers with an official-looking document authorizing him or her to collect funds and sign notes. The Service Socrates used a more complicated system, however. This organization invited prospective lenders to suggest a phrase to be mentioned on British Broadcasting Corporation (BBC) radio broadcasts on a given night. The underground passed the requests on to the London authorities, the phrase was broadcast at the designated time, and the individuals knew that they were dealing with bona fide agents of the underground. To safeguard the Service Socrates and the Belgian government in London against future false claims, lenders were given certificates stating the amount of the loans and bearing a number. Raymond Scheyven, using his pseudonym Socrates, signed these certificates, and a copy of this signature was on file in London for comparison at the time of repayment after the war.

If the underground can borrow in the name of some constituted authority such as a government-in-exile, it is more likely to receive a favorable response than if funds are sought in the name of an aspiring underground whose trustworthiness as a debtor organization may be in doubt. As one writer expressed it, governments-in-exile provide necessary "symbols of legalism."[18]

In addition, if an underground has access to some form of collateral, such as oil distribution networks or diamond fields, it may be able to secure funding, weapons, and other needed assets in exchange for granting access to this collateral or resource. The RUF in Sierra Leone obtained funds and weapons from Liberia and Libya in exchange for diamonds and access to mines.[19]

Embezzled Funds

An underground may obtain funds that have been embezzled from government agencies, trade unions, businesses, and nongovernmental organizations. An example is the secret appropriations that the Danish resistance received from the Royal Treasury to support the publication *Information*. Also, misappropriated grand duchy revenues constituted perhaps half of the money raised for the anti-Nazi resistance in Luxembourg. Trade union funds were embezzled on a fairly large scale by Communist leaders of Malayan trade unions in the years 1945–1947 and provided a major source of income for the Malayan Communist Party (MCP) until the British replaced the Communists with unionists loyal to the government. In Somalia, drought relief funding and supplies were interdicted by al-Shabaab to support its network, and in Rwanda, an estimated $112 million worth of foreign aid was used to purchase weapons—mostly machetes—from France, South Africa, and Egypt.[20]

Sales

The sale of various items through door-to-door canvasing or through "front" stores may provide money to resistance groups. Yugoslav Communists once sold fraudulent lottery tickets. The Luxembourg resistance sold lottery tickets as well as photographs of the grand duchess. In post-World War II Malaya, the MCP treasury was supplemented by funds obtained from party-owned bookstores, coffee shops, and even small general stores. Similarly, the Yugoslav Communists raised money through sales at party-owned clothing stores.

Coercive Means

Robberies

To bring in money, undergrounds frequently resort to robberies. The Hukbalahap in the Philippines, for instance, was able to collect funds by staging train robberies. Likewise, the Organisation de l'Armée Secrète (Organization of the Secret Army/Secret Armed Organization, or OAS) in Algeria conducted a series of bank robberies. In Malaya, the Communists formed a "Blood and Steel Corps" to engage in payroll robberies and raids on business establishments. Business firms, rather than individuals, are usually the targets of such robberies. The LTTE used numerous methods to mobilize resources to sustain operations. These methods included various criminal activities, such as bank robberies, extortion, and the smuggling of drugs and other contraband, but they also included more traditional fund-raising activities that also incorporated varying levels of coercion.[21]

Undergrounds generally avoid outright confiscations from the general populace for several reasons. First, widespread robberies tend to brand an underground as an outlaw band and destroy its public image as a potential legitimate authority. Second, simple confiscations of money are usually not as successful as other forms of coercion in making the victims compliant servants of the underground. Finally, robberies preclude the possibility of exacting continued support under the threat of exposing the affected persons' assistance to the underground.

Kidnapping and Hijacking

Kidnapping to collect a ransom has been a practice of insurgent groups around the world, including al-Nusra and Islamic State militants in Syria and Iraq, the Fuerzas Armadas Revolucionarias de Colombia (Revolutionary Armed Forces of Colombia, or FARC) in Colombia, the Taliban in Afghanistan, al-Qaeda in the Islamic Maghreb (AQIM) in Northern Africa, and the Movement for the Emancipation of the Niger Delta (MEND) in Nigeria. These organizations use elaborate networks of middlemen and negotiators to exchange their captives for funding. During the period of 2005–2010, AQIM alone raised an estimated $65 million from kidnappings, which accounted for 90 percent of its revenue and an average rate of $6.5 million for a Western hostage.[22] The dramatic escalation in hijackings off the Somali coast and into the Red Sea and Indian Ocean in 2008 to 2012 reportedly also lined the pockets of al-Shabaab insurgents who charged pirates and their villages a protection fee or tax after the ransom is paid.

Forced "Contribution"

Although undergrounds usually do not rob the public in an effort to avoid alienating their constituency, they sometimes coerce individuals into making donations under the tacit threat of reprisals. This technique is usually applied aggressively only to wealthier targets. The OAS in Algeria fixed the amounts of contributions to be exacted from persons in professional occupations but allowed people of modest means to give what they wanted. A target received a typewritten note in the mail informing him or her that a *percepteur* of the OAS would call in the near future to collect his or her contribution. *Percepteurs* were well dressed and curt but polite. If their credentials were questioned—some crooks tried to extort money in the name of the OAS—they could show a photostat message signed by the commander in chief of the OAS, General Raoul Salan. If the person refused to pay, he or she would not be threatened, but a week later his car or home would probably be bombed with plastic explosives. The OAS would then increase its assessment to cover the cost of the reprisal. After a few such object lessons, most of the people approached were willing to make a "contribution," and many agreed to make regular payments.

From its inception, the FARC in Colombia survived in part by obtaining funds through extortion, kidnapping and hostage-taking, and stealing supplies. At first, the FARC filled its coffers through extortion, using tactics that resembled those used by the Sicilian Mafia. The FARC asked businesses whether they needed "vaccinations." One Colombian Coca Cola distributor that decided it did not need to be inoculated suffered the FARC's response: forty-eight burned delivery trucks, eleven kidnapped workers (with seven eventually killed), and four hundred robberies.[23]

The Yugoslav Partisans were able to use this coercive technique to their political as well as financial advantage. By exacting large amounts from landowners, the Partisans were able to weaken their political opponents, and going one step further, they eliminated some of these political competitors by denouncing them to the Germans as helpers of the underground. A check of landowners' financial records sometimes revealed unaccountable deficits, which led to the landowners' arrest and the confiscation of their properties.

An underground may suffer a setback, however, if a popular person refuses to contribute and the underground does not dare to make the usual reprisal for fear of public indignation. A case in point is the widely publicized refusal of the French actress, Brigitte Bardot, to aid the OAS. Such a response serves to weaken the image of infallibility and complete control that the underground tries to cultivate.

Taxes

Taxes may be levied against the general public in areas where enemy forays are not frequent or serious enough to prevent underground municipal administrators from collecting taxes with the backing of nearby military units. Taxes may be levied on a per-capita basis, as was done in Philippine areas under Hukbalahap control, or they may be levied on a more selective basis, affecting only persons with regular incomes above a certain level, as was apparently the practice in the Slovene area of Yugoslav Partisan control.

The LTTE raised funds domestically by levying taxes on the population, especially in the early stages of the conflict. Those who could not pay were often incarcerated, while those families with sons or daughters serving in the cadres were exempt. Proof of payment of this tax served as a pass for traveling through LTTE-controlled territory and for serving in administrative positions. Extracting payment was relatively inexpensive; once the LTTE established its reputation for ruthlessness, few families had to be reprimanded.

For the Taliban, the decision to formalize and tax the heroin economy led to significant revenues for the insurgency. By 2000, Afghanistan produced three times more opium than the rest of the world combined, with 96 percent of this Afghan production taking place in Taliban-controlled territory. The Taliban collected a 20 percent tax from opium dealers and drug transporters, leading to an annual tax revenue base of approximately $20 million by 2000.[24]

Narcotics and Black Market Trade

Modern insurgencies have increasing connections to illegal drug trade throughout the world. The burgeoning industry of supplying marijuana, cocaine, methamphetamine, heroin, and other drugs offers opportunities for financing that most underground leaders find too lucrative to ignore.

Although extortion and kidnapping sustained the FARC for many years, growing the "little guerrilla army" required a corresponding growth in funding. To do this, during the Seventh Guerrilla Conference, the FARC developed a plan to leverage four commodities on the black market: livestock, commercial agriculture, oil, and gold. This revenue would not prove to be enough, however. Eventually, without departing from the extortion and kidnapping, FARC reluctantly became involved in the narcotics trade. Initially, both Manuel Marulanda and Jacobo Arenas were opposed to *las drogas* for ideological reasons. In the long run, however, pragmatism won. The flow of illicit money and goods through FARC-controlled areas was just too rich a source to ignore. In 1998 alone, for example, armed groups involved in the marketing of illegal narcotics in Colombia saw proceeds in excess of $550 million. Still, the FARC became involved piecemeal. Its first step was to tax narcotraffickers while protecting the peasant farmers who grew the coca, a practice that was not unlike its "inoculations."[25]

The estimates of financing FARC obtained through narcotics range from at least $30 million annually to as high as $1.5 billion. In fact, the FARC became so sophisticated that it developed standard costs for the drug trade that in October 1999 equated to $15.70/kilo for cocaine paste, $5,263 to protect a laboratory, and $52.60 to protect a hectare of coca. At one point, the FARC was responsible for exporting 50 percent of the cocaine consumed worldwide. Whatever the figure, the revenues from the narcotics trade still represented only half of the FARC's funding. The Colombian government suggests that the rest came from the FARC's classic funding lines of kidnapping, robbery, and extortion.[26]

Involvement in the drug trade or other black market activities often impacts the organization's core ideology and strategy. The FARC experience, again, is illustrative. With the new drug trade came money and corruption. Some FARC members in coca-rich areas began to live as drug lords, replete with gold jewelry, fancy cars, and other luxuries. This created dissent in the ranks as FARC members who stayed true to the guerrilla life realized others were living as gangsters. To solve this problem, the FARC leadership created the National Financial Commission. Responsible for allocation of all FARC funds, including major purchases, the commission reported directly to the central leadership. A system was developed wherein all FARC units were given a funding line and direction in how to use it. When these measures did not completely solve the corruption issue, the senior leaders assigned *ayudantías*, or advisers, to monitor what was happening at every level and provide advice to local leaders from time to time. If they suspected any foul play, the *ayudantías* would call for an investigation. Theft or even misappropriation of FARC funding was punishable by death.[27]

Sendero Luminoso, the Maoist insurgency in Peru, likewise became involved in the illegal drug trade. By the mid-1980s, having expanded in the primary coca-producing region of Peru, the Upper Huallaga Valley, the Sendero Luminoso, or the Shining Path, was able to tap the profits of the drug traffickers. To exploit the ever-burgeoning cocaine trade, the group sent units into the Upper Huallaga Valley to identify and kill government enforcement agents and their supporters. Once they had taken effective control of the region, Sendero members served as middlemen between the coca growers and the drug traffickers, reportedly receiving 10 percent of the sale of every kilo of coca and earning between $20 and $50 million annually, which was used to purchase weapons and pay militants. Although Abimael Guzman originally disavowed any connection between Sendero Luminoso and the ongoing drug trafficking along the Andean ridge, pragmatic considerations appear to have triumphed. However, the perceived disconnect between Sendero's purported ideological purity and its involvement in the black market drug trade did impact its legitimacy.

TACTICAL ECONOMIC ANALYSIS

Obtaining Operational Advantage via Understanding and Influencing the Adversary's Economic Behavior

"Amateurs talk strategy, professionals talk logistics."
—General Omar Bradley

From sanctions to development programs, economic and financial instruments of national power are crucial components of military and interagency operations, strategy, and planning, but they are often underused or misused. The behaviors, capabilities, and ultimate success or failure of an adversary—or indeed also an ally—are closely tied to and sometimes determined outright by economic and financial factors. These factors not only *impact* an actor's conduct and capabilities but also *reflect* their social, political, and operational relationships and dynamics. This reality makes adversaries vulnerable to disruption, destruction, or manipulation of their economic and financial relationships, access points, and supply lines and also opens up potentially important new subsets of exploitable information against or in support of a given actor. Economic and financial instruments are thus, as current joint doctrine observes, some of "the most powerful ideas in the arsenal of US power"[28]—a reality also reflected in the emerging interagency focus on "counter-threat finance" (CTF), as well as anti-trafficking, anti-money laundering, and other distinct but related efforts.

The problem, however, is that despite a consensus across the US government and among allies that understanding and influencing the economic domain is important—even to the extent of waging a so-called financial war on terrorist threats—to date this potential has gone largely unrealized. For example, military efforts to track and disrupt the financial bases of the insurgencies in Iraq and Afghanistan remain unsystematic and ad hoc, with some notable successes but also wasted effort and unrealized potential.[29] Efforts of civilian agencies are similarly disconnected from both complementary programs within other agencies and even the economic realities of the threats they are meant to counter, particularly when it comes to understanding and exploiting adversaries' economic activity outside of monetary-based Western formal financial systems. Overall, this lack of coordination in understanding the economic bases of our adversaries has resulted in a situation in which, in the words of knowledgeable observers, we are, "looking in the wrong places,"[30] "asking the wrong questions,"[31] and "fighting terror with error"[32] to such an extent even that "our enemies are laughing at us."[33]

Partly, this is because economic and financial realities inherently resist simple definitions and categories, narrowly conceived and separated bureaucratic responsibilities, or direct and simplistic application of power, especially in societies outside of North America and Western Europe.[d] It also requires in a sense "knowing everything about everything," in that successful use of economic and financial instruments depends on collecting and exploiting—correctly—often highly localized and tactical information, even though the impact of these instruments can be strategic and global in effect.[34] But even more significantly, this failure relates to the lack of a holistic, systematic, "big-picture" methodology for integrating and using for effect relevant information and practice—most of which, this paper argues, already exist but exist at local and tactical levels. Without such a methodology, use of economic and financial instruments will continue to have, at best, only ad hoc and sporadic impact, at least in terms of the ability to actually achieve national objectives rather than to simply justify organizational budgets and missions.

Overview of the Tactical Economic Analysis Methodology

Tactical economic analysis (TEA) is a holistic, systematic, big-picture methodology for integrating and using for effect information and practice relevant to economic and financial instruments of national power, initially within the Department of Defense but also eventually across the US government and among US allies as well as within the private and nongovernmental sectors.

In summary, the methodology applies the F3EAD (find, fix, finish, exploit, analyze, disseminate) focused targeting method to the economic and financial domain. The TEA method starts with two parallel baseline assessments of the target adversary's critical economic and financial requirements as well as of the local economic and human terrain with which the actor must interact to meet these requirements. From this, one can then identify potential targetable vulnerabilities via investigation of particular gaps, overlaps, linkages, and anomalies between these two baselines, and in turn develop potential courses of action, assess them (for impact, risks, and deconfliction of effort), and ultimately decide on, execute, and exploit for advantage one or more of the vulnerabilities.

The TEA method was developed in 2011 by the political economist Timothy Wittig in partnership with the US Army Asymmetric Warfare

---

[d] This is a paraphrase of FM 3-05.130's description of some of the inherent difficulties of unconventional warfare—phrasing that fits here as well.

Group in Fort Meade, Maryland, and later refined with input or funding from the Naval Special Warfare Development Group; the US Special Operations Command CTF detachments at US Pacific Command, US Africa Command (USAFRICOM), and US Central Command; the Johns Hopkins University Applied Physics Laboratory; and numerous other military, civilian, and academic partners.

Focused targeting is a process for "developing actionable intelligence and hitting targets" in a way that "maintains the initiative" against an adversary. Although systematic and iterative, focused targeting is as much art as science, and a key factor for success is the methodical and continuous analysis and exploitation of ground intelligence "to conduct immediate follow-on operations."[35] TEA offers a means to achieve similar successes within and via the economic domain and thus represents a potentially important contribution to the CTF, counterterrorism, counterinsurgency, unconventional warfare, foreign internal defense, civil affairs, and counternarcotics mission sets, especially pertaining to nonlethal targeting, and improving collection of human, communications, and open-source intelligence.

The primary innovation of the TEA methodology is its shift of analytic focus from individuals or networks (which are often easily replaced or reconstituted) to "value chains" and the threat's interaction with them. Targeting threat actors' access to that which they most value enables more holistic approaches to countering, undermining, and defeating a variety of transnational threats to the United States. In particular, TEA makes the following general contributions to existing military practice:

- A common lexicon for the economic and financial dimensions of adversary activity
- A method for systematically identifying adversary vulnerabilities (economic and noneconomic)
- A method for identifying and evaluating potential courses of action against these vulnerabilities
- A framework for prioritizing adversary-related economic and financial information and intelligence that is applicable across different mission sets, geographic locations, and operational contexts
- A framework for linking existing information and practice, in particular CTF, anti-trafficking, and anti-money laundering programs with wider political, diplomatic, military, law enforcement, intelligence, and foreign aid efforts
- A structure with which policy, action, training, and acquisition for economic and financial instruments of national power can be made more accountable and can be better assessed and evaluated

Overall, the TEA method dispenses with the pretense of being the "next big thing" and instead emphasizes above all better integration and use of existing information and practice to gain advantage against an adversary via targeted exploitation of its economic dimensions. The following sections outline the concepts, tools, and informational requirements of each component of this proposed methodology, along with how these apply to the case of al-Shabaab in Somalia.

Establishing a Baseline: The Analyze Phase

The analyze phase is the first and in many ways most important phase of the TEA method. It consists of two parallel informational baselines—(1) the human and socioeconomic terrain in which the adversary operates and (2) the adversary's material and nonmaterial needs and how these are met. The focus for each is on *initial* rather than comprehensive observation that is conducted within one's own resource, authority, and other constraints. That said, the unavoidable fact is that the better and more comprehensive these baselines are, the more efficient and more effective the later targeting will become.

Using PMESII-ASCOPE for Threat Finance Exploitation

One baseline is an assessment of the socioeconomic terrain, or in other words the licit and illicit markets, goods, services, and value exchange systems in a given area of operations, as well as the social, political, and economic structures that shape and are shaped by them. The information requirements for the socioeconomic terrain assessment are, in general terms, the same as those required for a PMESII-ASCOPE assessment (see Table 1 for a sample PMESII-ASCOPE table).

The emphasis is on collection and aggregation of information about both specific entities (e.g., people, businesses, and locations) as well as value chains (i.e., the ways in which these entities interact with one another to produce or trade economic value). The concept of the value chain is important as it explains how various PMESII-ASCOPE factors relate to and interact with one another for the economic advantage or disadvantage to an adversary or ally. Various definitions of value chain exist,[e] but the term is most simply understood as the flows of economic

---

[e] The term *value chain* is a useful but somewhat unsettled concept within academic literature, defined variously yet similarly as: "the shifting governance structures in sectors producing for global markets;" "the process by which technology is combined with material and labor inputs, and then processed inputs are assembled, marketed, and distributed;" and "the sequential set of primary and support activities that an enterprise performs to turn inputs into value-added outputs for its external customers." For an extended discussion of the value chains in context of contemporary threats, see Wittig.[36]

value that circulate from, to, or through a particular time and place. The key innovation of this methodology is to holistically (and typically nonlethally) target enemy interaction with these flows, rather than simply target the entities of which they (often only temporarily) consist. The details of this interaction will vary widely over time and place, but for reference the following list specifies which PMESII-ASCOPE elements are most commonly useful for this baseline analyze phase assessment of the relevant socioeconomic terrain:

- Aid, charity, religious, and educational entities
- Aid, charity, religious, and educational value chains
- Commercial practices and customs
- Financial services, entities
- Financial services, value chains
- Goods traded
- Goods, value chains
- Governmental organizations, entities
- Governmental organizations, value chains
- Laws/regulations/related authorities
- Productive enterprises, value chains
- Productive enterprises, entities
- Real properties
- Real property, value chains
- Services available
- Sociopolitical movements
- Souks/bazaars/marketplaces, entities
- Souks/bazaars/marketplaces, value chains
- Taxes/tariffs/checkpoints, entities
- Taxes/tariffs/checkpoints, value chains
- Trade/trafficking routes, entities
- Trade/trafficking routes, value chains
- Transportation systems, entities
- Transportation systems, value chains

Table 1. Sample PMESII-ASCOPE table.

	P Political	M Military/ Security	E Economic	S Social	I Infrastructure	I Information
A Areas	District / Provincial Boundary	Improvised Explosive Device (IED) Sites, Military/ Insurgent Bases	Bazaars, Farms, Repair Shops	Picnic Areas, Bazaars, Meeting Sites	Irrigation Networks, Medical Services	Radio, Gathering Points, Graffiti, Posters
S Structures	Shura Halls, Court House	Police Headquarters, Military Bases	Bazaars, Banks, Industrial Plants	Mosques, Wedding Halls	Roads, Bridges, Electrical Lines, Dams	Cell/Radio/TV Towers, Print Shops
C Capabilities	Dispute Resolution, Judges, Local Leadership	Military/Police Enemy Recruiting Potential?	Access to Banks, Development, Black Market	Traditional Structures, Means of Justice	Ability to Build / Maintain Roads, Dams, Irrigation	Literacy Rate, Phone Service
O Organizations	Governmental organizations, NGOs	Coalition and Host Nation Forces	Banks, Landholders, Economic NGOs	Tribes, Clans, Families	Governmental Ministries, Construction Companies	News Organizations, Mosques
P People	Governors, Councils, Elders, Judges	Coalition/Host Nation Military/ Police Leaders	Bankers, Landholders, Merchants, Hawaladars, Criminals	Religious/Civic Leaders, Elders, Families	Builders, Contractors, Development Councils	Civic/Religious Leaders, Family Heads
E Events	Elections, Meetings, Speeches, Trials	Kinetic Events, Military/Police Operations	Drought Harvest, Business Opening	Weddings/ Deaths, Births, Funerals, Bazaar Days	Road/Bridges, School Construction, Well Digging	Festivals, Project Openings

In practice, economically relevant PMESII-ASCOPE information is just as likely to be highly specific and localized as it is to be general and broad, and it will be of uncertain significance until the find and fix phases of the TEA method are completed. For instance, a study conducted for the Naval Special Warfare Development Group and National Defense University on al-Shabaab's financing between 2009 and 2012, when the group was the de facto government in both Somalia's capital Mogadishu and the southern half of the country, found that one of al-Shabaab's key exploitable vulnerabilities was its relationship with the lucrative sugar trade through southern Somalia's ports. This finding was based on, among other things, the following PMESII-ASCOPE information:

- An estimated twenty thousand to forty thousand metric tons of sugar were imported each year into southern Somali ports, primarily Kismaayo but also Marka, Baraawe, and Buur Gaabo Eel Ma'an Qudha.

- Al-Shabaab effectively governed these ports and controls the territory around them, meaning that the group is able to earn considerable revenues—an estimated $7 to $10 million per year, or 10 percent of its annual revenue, according to the United Nations (UN)—from taxing the sugar trade moving through these ports.

- The bulk of this sugar was shipped in fifty-kilogram sacks into Somalia by import-export businesses based—and legally registered— in the United Arab Emirates (UAE), and it was then smuggled overland into Kenya for resale.

- In Kismaayo, for example, the import duty for sugar and other foodstuffs was, as of April 2011, $1.00 per fifty kilograms—a rate set by the Kismaayo port authorities who by necessity must abide by what al-Shabaab dictates and which doubled compared to winter 2011.

- This lucrative value chain appeared to be dominated by just two UAE-based companies—Nour Mowafaq General Trading LLC and Sinwan General Trading LLC—with which al-Shabaab has apparently established mutually beneficial business relationships.

- Sugar imports in Somalia are historically closely linked to the export trade in charcoal via the same ports, meaning that it is likely that targeting the sugar import trade will help eliminate the charcoal trade as well.

Although the tools available to obtain this initial observation of the economic terrain features will vary in accordance with the resources available in a specific area of operations, in general they will include existing or custom-tasked intelligence, surveillance, and reconnaissance and atmospherics reporting; relevant civilian information products (e.g., economic development data); academic publications; media reporting; and interviews (formal and informal) with local civilians, researchers, and other knowledgeable parties. USAFRICOM, for example, has a number of programs that could, hypothetically at least, provide economic terrain information, including the command's Social Science Research Branch Socio-Cultural Analysis Teams, which has a deployed element in Djibouti, and the various ad hoc partnerships and exchanges USAFRICOM and other relevant US government entities (such as National Defense University's Africa Center for Strategic Studies) regularly conduct with universities and individual external researchers. In the context of the al-Shabaab case study, economic terrain information was obtained from the following sources:

- Interviews (formal and informal) with members of the Somali diaspora in Nairobi, Djibouti, and Dubai; employees of humanitarian NGOs working in Somalia; US government officials, primarily from the US Agency for International Development and the Departments of State, Defense, and Justice; UN officials; and researchers and journalists working in or reporting on Somalia

- Personal observation, especially of business and trade dynamics in Nairobi, Djibouti, and Dubai

- Published academic research and media reporting, especially economic anthropology literature that details the ethnography and technical processes of trade and business in and via Somalia and within Somali diaspora communities
- Gray literature, such as unreleased internal reports from relevant organizations and other unpublished material

Regardless of the specific mixture of tools used, it is important to note that virtually all of this economic terrain information is already being collected by someone, but may not be—and indeed probably is not being—applied to relevant efforts or missions.

The specific outputs of an economic terrain program will vary depending on resources and organizational priorities but in general will include mechanisms and authorities to request and collect relevant information (such as internal taskings or requests for information from external partners), as well as some kind of repository for this information (such as a database, a series of relevant and appropriately updated reports, maps and graphics, or even simply an individual tasked to this purpose).

Overall, the economic terrain baseline requires a focus on initial rather than comprehensive collection and analysis (any gaps or deficiencies that exist can and will be filled in later as needed), and given that most of the required information is likely already accessible, commitment to identifying reliable and high-quality sources of economic terrain information regardless of whether they are military or non-military, classified or open source, or friendly or not.

Adversary Critical Resource Assessment

The other baseline to be conducted in parallel as part of the analyze phase is referred to as the adversary critical resource assessment. The critical resource baseline also emphasizes initial rather than comprehensive or definitive observation of known threat actors within an area of operations and the goods, services, and economic systems they use or control. When combined with economic terrain assessment, this is analogous to persistent surveillance of the economic bases of the enemy. The information requirements for the adversary critical resource assessment are outlined in Table 2.

As many readers will immediately note, much of this information—economic and noneconomic—is already collected and aggregated via established channels (such as those related to the specially designated nationals lists, high-value individuals lists, joint effect list, or Joint prioritized effects list), as well as by various other analysts including other own-government offices (e.g., economic development), academics and researchers, and allied governments. In the case of the al-Shabaab study, similar sources were used, although with perhaps a heavier reliance on interviews with NGO and governmental officials and the gray literature.

As with the parallel economic terrain assessment described above, the desired end state is not a final, comprehensive picture but simply an initial observation of the enemy's economic requirements, activities, relationships, and dynamics to serve as a baseline from which systematic, targeted analysis can be conducted. As such, the tangible end products for this phase may include, depending on one's resources and organizational priorities, intelligence requirements for field and headquarters elements; related databases, reports, and bulletins; and reach-back support from other agencies or allied governments.

Table 2. Information requirements for adversary critical resource assessments (for each threat actor present).

Background attributes:
- Demographic composition (ethnic, sex, age, religion, etc.)
- Ideologies (explicit and implicit)
- Objectives, strategies, and tactics
  - Military
  - Political
  - Socioeconomic
  - Internal organizational
- Internal structures, processes, and culture (e.g., hierarchies, networks, discipline, dispute resolution, etc.)
- Personalities
- Leadership
- Other key personnel
- External relationships

Critical resources and resource requirements:
- Financial resources
- Human resources
- Logistical resources
- Military resources
- Sociocultural resources
- Political resources

Economic interaction with:
- Aid, charity, religious, and educational entities
- Financial services entities
- Governmental organizations entities

Economic interaction with (continued):
- Productive enterprises entities
- Souks/bazaars/marketplaces entities
- Taxes/tariffs/checkpoints entities
- Trade/trafficking routes entities
- Transportation systems entities

Social interaction with:
- Aid, charity, religious, and educational entities
- Financial services entities
- Governmental organizations entities
- Productive enterprises entities
- Souks/bazaars/marketplaces entities
- Taxes/tariffs/checkpoints entities
- Trade/trafficking routes entities
- Transportation systems entities

Access to:
- Aid, charity, religious, and educational institutions, value/supply chains of
- Financial services industry
- Goods, trade flows in
- Governmental organizations, value/supply chains of
- Productive enterprise, value/supply chains of
- Real property market

Access to (continued):

- Souks/bazaars/marketplaces, value/supply chains of
- Taxes/tariffs/checkpoints, value/supply chains of
- Trade/trafficking routes, value/supply chains of
- Transportation systems, value/supply chains of

Influence on:

- Aid, charity, religious, and educational institutions, value/supply chains of
- Financial Services industry
- Goods, trade flows in
- Governmental organizations, value/supply chains of
- Productive enterprise, value/supply chains of
- Real property market
- Souks/bazaars/marketplaces, value/supply chains of
- Taxes/tariffs/checkpoints, value/supply chains of
- Trade/trafficking routes, value/supply chains of
- Transportation systems, value/supply chains of

Servicing of:

- Aid, charity, religious, and educational institutions, value/supply chains of
- Financial services industry
- Goods, trade flows in
- Governmental organizations, value/supply chains of
- Productive enterprise, value/supply chains of
- Real property market
- Souks/bazaars/marketplaces, value/supply chains of
- Taxes/tariffs/checkpoints, value/supply chains of
- Trade/trafficking routes, value/supply chains of
- Transportation systems, value/supply chains of

Material support to and from:

- Material support to adversary's nonviolent operations
- Material support to adversary's violent operations
- Material support to sociopolitical movements related to adversary

Individual transactions:

- Locations
- Entities involve
- Date-Time Group
- Relevant linkages

One ancillary benefit of the framework presented here is that it provides a home for pieces of data that appear relevant but the actual usefulness of which may be unclear. For instance, the al-Shabaab case study exposed the following data points about the group's critical resource requirements, among thousands of others:

- **Financial resources:** A firefight with African Union Mission to Somalia (AMISOM) forces in Mogadishu costs al-Shabaab approximately $10,000 per hour (before the group's withdrawal from the city in July 2011).

- **Human resources:** Al-Shabaab uses *adoosi* as conscript fighters. (*Adoosi*, the Somali word for slave, typically refers to ethnically Bantu Somalis who historically have been exploited by militias to fight in exchange for being able to loot civilian villages.)

- **Sociocultural resources:** Al-Shabaab appears to be a fully covert organization outside of Somalia, including in Eastleigh, Nairobi, the predominantly Somali area of the city and where media reporting had indicated the group operates openly.

Prioritizing Finite Analytic Resources: The Find Phase

The second phase, the find phase, of the TEA methodology aims to identify and investigate points of information that may indicate potentially exploitable vulnerabilities in the target group's economic, social, political, military-operational, or other behavior. Most simply, this information is uncovered by comparing the two baselines from the analyze phase to identify and then investigate any evident gaps, overlaps, or anomalies that may indicate exploitable interaction by the enemy with known economic terrain.

The analyze phase by design is a clearinghouse for the huge amount of information about the adversary and its socioeconomic contexts. The find phase, by contrast, aims to identify only the highest priority sets of information—the suspicious activities or red flags—against which an organization's finite analytic resources can be tasked. The intended end state of this phase therefore is a well-planned and well-executed investigation of each point of suspicious activity to determine whether it represents a vulnerability that could possibly be exploited for advantage in some way. When it does, then courses of action against this vulnerability will be developed in the following fix phase of the method. However, often such investigations will result in a determination that the identified suspicious activity should not be promoted to the fix phase, because for example it had an innocent (or at least nonthreatening) explanation or perhaps simply a lack of corroborating evidence. Even in these cases, the TEA method helps ensure that no effort is wasted, in that even unsuccessful investigations will necessarily result in improved analyze phase baselines, which will be useful in future cases.

In the Somalia study, dozens of potential vulnerabilities in al-Shabaab were identified using this method, which for the purposes of brevity can be aggregated into seven major vulnerabilities:

1. **Sugar imports:** Al-Shabaab gains significant revenues from taxing the importation of sugar into Kismaayo and other southern Somali ports, a trade that is dominated by only two UAE-based companies with which al-Shabaab is believed to have close, but primarily transactional, relationships. By targeting these particular relationships, a major source of revenue for the group could be disrupted if not eliminated outright.

2. **Charcoal exports:** Similar to the sugar trade, al-Shabaab taxes the exportation of charcoal produced in Somalia to primarily the Gulf states, via privileged business relationships with a small number of UAE-based shipping companies. This vulnerability is similar to the vulnerability related to the sugar trade.

3. **Air traffic:** Al-Shabaab relies on regular (scheduled and chartered) air transport links between southern Somalia and nearby countries for transport of cash, supplies, and personnel, most of which is legal or quasi-legal. Systematic monitoring of air traffic in and through al-Shabaab-controlled areas could yield important tactical-level intelligence about the group.

4. **Material supporters:** Al-Shabaab obtains significant financial and logistical support from both state and non-state supporters, ranging from cash donations from wealthy sympathizers to logistical and technological support from state governments. This support is in many ways shallow and transactional, depending completely on al-Shabaab continuing to be a proxy through which its supporters can meet their varied objectives. These support flows could be diminished or eliminated by changing perceptions of al-Shabaab and/or the group's ability to act as a proxy.

5. **Bakaara market:** Although al-Shabaab has long profited from business relationships with the arms and other merchants of Mogadishu's Bakaara market, these relationships are highly dependent on al-Shabaab maintaining territorial control not just of Mogadishu but also of the city's suburbs where many of Bakaara's merchants keep their inventories. Continuing territorial gains from al-Shabaab in and around Mogadishu would be disproportionately more significant than elsewhere.

6. **Humanitarian aid flows:** Although al-Shabaab obtains significant revenues from taxing and diverting humanitarian aid, these flows at the same time cause significant internal dissension and even discord among the various factions within the group, caused by differing ideologies about accepting aid from Westerners and other "enemies," even in the face of widespread suffering among the populations under al-Shabaab control. These divisions could be documented and exploited for advantage against al-Shabaab.

7. **Livestock, telecommunications, and *xawilaad* industries:** Interestingly, al-Shabaab barely interacts with these three industries—the Somali economy's most important and profitable—and when it does, it is often in an antagonistic way. Somalia's powerful business elite is likely to be either indifferent or even actively disposed to oppose to al-Shabaab.

Each of these findings indicates a potential vulnerability in al-Shabaab, but determining how each can be best exploited is the subject of the next phase of the methodology.

Target Development and Risk-Benefit Assessment: The Fix Phase

The fix phase is characterized by the systematic fusion, analysis, and dissemination of information involving the value chains accessed by threat actors until a specific actionable point of exploitation can be predicted against each previously identified vulnerability.

The fix phase has three subsidiary steps. The first is to confirm or deny assumptions regarding value chain access points involving threat actors. This is to provide a reality check and ensure that the identified target actually does represent a potentially exploitable vulnerability. This is accomplished by constructively challenging each and every assumption on which this determination is based, including both those pertaining to the adversary itself as well as the relevant economic terrain. Second, it is necessary to develop a range of courses of action that at least are theoretically available against each vulnerability identified in the find phase. These courses of action will generally fall into one of four categories of actions:

1. Shaping actions (i.e., actions that aim to limit the range of options the object has for action)
2. Disruption actions (i.e., any action meant to interrupt or otherwise disrupt the adversary's interaction with the targeted value chain)
3. Kinetic actions
4. Monitoring actions

In the case of al-Shabaab, a number of courses of action are possible, including, for instance, those listed in Table 3.

Table 3. Abridged outline of a potential coordinated campaign against the economic foundations of al-Shabaab.

Shaping actions:

- Designate all trade via Kismaayo, Marka, Baraawe, and Buur Gaabo Eel Ma'an Qudha ports as violation of relevant UN Security Council Resolutions and US law.
- Subsidize US charcoal exports to Gulf Cooperation Council countries.
- Engage with and/or create and exploit leverage against Somali business elite, especially in the livestock, telecommunications, and finance (*xawilaad*) industries.
- Provide capacity-building support to major livestock middlemen.
- Lower duties in Mogadishu ports.
- Enter into discreet negotiations with local al-Shabaab commanders.
- Conduct relevant information operations.

Kinetic actions, against:

- Beach ports in al-Shabaab-controlled areas of Somalia
- Bakaara arms market inventories
- Al-Shabaab *shura*
- Key foreign fighters

Disruption actions:

- Actively encourage humanitarian aid flows into al-Shabaab-controlled areas.
- Make the export of charcoal illegal in Somalia.
- Create and exploit leverage against material supporters of al-Shabaab to get them to stop or diminish donations to the group.
- Create and exploit leverage against those individuals and companies that do business with al-Shabaab or in al-Shabaab-controlled areas.

Monitoring actions, focused on:

- Bakaara market inventories
- Companies that import goods into al-Shabaab-controlled locations
- Companies that export goods from al-Shabaab-controlled locations
- Somali business elite, inside and outside Somalia
- Al-Shabaab factional divisions
- Air traffic into, out of, and within southern Somalia
- Trade between Somalia and ports in Yemen, Djibouti, and the UAE

The third and final step of the fix phase is a comprehensive assessment of the PMESII impacts of each exploitation option, both positive and negative, namely the following:

- Operational benefits and risks, such as to friendly forces and risk of confliction with other efforts
- Political benefits and risks, such as to strategic policy or particular political or diplomatic relationships
- Social benefits and risks, such as relating to intercommunal or intracommunal relations among civilian populations

- Economic benefits and risks relating to both the resource costs of the action, as well as the economic welfare of friendly nations and civilian populations
- Informational benefits and risks relating to both friendly and adversary information narratives

In the al-Shabaab case, for example, kinetic action against the port of Kismaayo would have severe and lasting economic effects on Somali civilians. However a more creative approach might include a combination of actions such as creating legal leverage via changes in international and US law combined with various diplomatic, intelligence, capacity-building, and even kinetic action.

The tangible end result of the fix phase is what could be called a target folder, including a recommended course of action against a specific target, as well as a comprehensive paper trail that includes the justification for this target as well as a rigorous examination of the potential second-, third-, and fourth-echelon effects of any potential course of action.

Taking Action, Gaining Advantage: Finish, Exploit, and Disseminate Phases

The final phases of the methodology rely on established practices to execute recommendations from the targeting package developed in the find phase. In practical terms, the biggest challenge is obtaining buy-in from the executing arms.

The goal of these phases is to successfully target the vulnerability and then gain as much advantage from doing so. Strategically, successful targeting of an adversary's economic foundations will generally either negatively impact the adversary's operational, social, political, economic, or other capabilities; facilitate a better understanding of the adversary's behaviors, relationships, trends, strategies; or both. Although more work is required to confirm this, it can be assumed that most advantage will actually be at the tactical level, such as through the following:

- Disrupting enemy supply lines for needed equipment/people/ services
- Creating windows of opportunity and/or exploitable pressure points
- Degrading specific operational capabilities
- Disrupting the enemy's exploitation of the local economy
- Reducing the enemy's disruption of development projects

- Strengthening the local population's independence from enemy-connected value chains
- Undermining the enemy's relationships with patrons, proxies, and the population
- Discrediting the enemy's objectives, and operations
- Creating or validating advantageous information operations narratives

In the case of al-Shabaab, its critical resource needs are met via a known and finite set of value chains. Importantly, al-Shabaab's primary sources of revenue are not crucial to the stability of Somalia's society or economy. This is a significant finding with both strategic and tactical importance, as it implies that targeting al-Shabaab's access to these flows of economic value could degrade its capabilities and change its behavior but could do so without exacerbating existing security and humanitarian problems. This latter point is crucial, especially in the context of the many well-intentioned but often counterproductive actions of the United States and the international community in Somalia over the last twenty years. When combined with other actions (kinetic, information operations, sociopolitical), this would likely lead to the defeat of al-Shabaab and would open the door for meaningful stability in Somalia.

Overall, economic and financial instruments of national power are also critically important elements of US operations and policy. Use of these instruments, however, remains largely ad hoc, poorly conceptualized, and unsystematically integrated into other priorities and efforts, especially at tactical levels for SOF. It is important to improve the use of economic and financial instruments of national power to systematically exploit economic and financial information and to target a given adversary's economic and financial dynamics at tactical levels. In particular, the TEA methodology does the following:

- Provides a more systematic approach to analysis and exploitation of the economic domain
- Improves data structuring and organization
- Improves the prioritization of finite information and action resources
- Makes analysis and action efforts more accountable
- Exposes new subsets of information
- Exposes new exploitable vulnerabilities in threat actors
- Enables coordinated economic-based campaigns against threat actors
- Enables better risk management

CONCLUSION

Financing is the lifeblood of underground movements. Economic activity both impacts the behavior of militant groups and reflects key information and insight into an adversary. Systematically understanding and leveraging for advantage these economic dimensions is a critical element of successful SOF mission accomplishment against terrorist, insurgent, and transnational criminal actors.

NOTES

[1] Financial Action Task Force/Groupe d'action financiere, "Terrorist Financing" (Paris: OECD, 2008), 7–10, http://www.fatf-gafi.org/dataoecd/28/43/40285899.pdf.

[2] Andrew R. Molnar, *Undergrounds in Insurgent, Revolutionary, and Resistance Warfare* (Washington, DC: Special Operations Research Office, American University, 1963), 62.

[3] Ibid.

[4] Maegen Nix and Shana Marshall, "Liberation Tigers of Tamil Eelam (LTTE)," in *Casebook on Insurgency and Revolutionary Warfare, Volume II: 1962–2009*, ed. Chuck Crossett (Fort Bragg, NC: United States Army Special Operations Command, 2009), 233–275.

[5] Ron Buikema and Matt Burger, "New People's Army (NPA)," in *Casebook on Insurgency and Revolutionary Warfare, Volume II: 1962–2009*, 5–38.

[6] Chuck Crossett and Summer Newton, "The Provisional Irish Republican Army: 1969–1998," in *Casebook on Insurgency and Revolutionary Warfare, Volume II: 1962–2009*, 379–421.

[7] Financial Action Task Force, "Terrorist Financing," 11–12.

[8] Molnar, *Undergrounds*.

[9] Darlene Storm, "Intelligence Agencies Hunting for Terrorists in World of Warcraft," *ComputerWorld* (blog), April 13, 2011, http://blogs.computerworld.com/18131/intelligence_agencies_hunting_for_terrorists_in_world_of_warcraft; and "Al-Shabab Bans Mobile Phone Money Transfers in Somalia," *BBC*, October 18, 2010, http://www.bbc.co.uk/news/world-africa-11566247.

[10] Rachel Ehrenfeld, "The Muslim Brotherhood New International Economic Order," *The Terror Finance Blog* (blog), October 13, 2007, http://www.terrorfinance.org/the_terror_finance_blog/2007/10/the-muslim-brot-1.html.

[11] Ibid.

[12] Ibid.

[13] Lawrence Wright, *The Looming Tower: Al-Qaeda and the Road to 9/11* (New York: Vintage Books, 2006).

[14] Peter L. Bergen, *Holy War, Inc.: Inside the Secret World of Osama Bin Laden* (New York: The Free Press, 2001).

[15] Michael Scheuer, *Through Our Enemies' Eyes: Osama Bin Laden, Radical Islam, and the Future of America* (Washington, DC: Potomac Books, Inc., 2008).

[16] Marc Sageman, *Understanding Terrorist Networks* (Philadelphia: University of Pennsylvania Press, 2004).

[17] Mohammed El-Qorchi, "The Hawala System," *Finance and Development* 39, no. 4 (December 2002), http://www.gdrc.org/icm/hawala.html.

[18] Molnar, *Undergrounds*, 64.

19 Jerome Conley, "Revolutionary United Front (RUF)—Sierra Leone," in *Casebook on Insurgency and Revolutionary Warfare, Volume II: 1962–2009,* 763–800.

20 Bryan Gervais, "Hutu–Tutsi Genocides," in *Casebook on Insurgency and Revolutionary Warfare, Volume II: 1962–2009,* 307–342.

21 Nix and Marhsall, "Liberation Tigers."

22 Jean-Charles Brisard, "AQIM Kidnap-for-Ransom Practice," *The Terror Finance Blog* (blog), September 27, 2010, http://www.terrorfinance.org/the_terror_finance_blog/2010/09/aqim-kidnap-for-ransom-practice-a-worrisome-challenge-to-the-war-against-terrorism-financing.html.

23 Ron Buikema and Matt Burger, "Fuerzas Armadas Revolucionarias De Colombia (FARC)," in *Casebook on Insurgency and Revolutionary Warfare, Volume II: 1962–2009,* 39–70.

24 Sanaz Mirzaei, "Taliban: 1994–2009," in *Casebook on Insurgency and Revolutionary Warfare, Volume II: 1962–2009,* 651–684.

25 Buikema and Burger, "FARC."

26 Ibid.

27 Ibid.

28 Field Manual 3-05.130 (FM 3-05.130), *Unconventional Warfare* (Washington, DC: Headquarters, Department of the Army, September 2008).

29 Gretchen Peters, *Seeds of Terror: How Drugs, Thugs, and Crime Are Reshaping the Afghan War* (New York: St. Martin's Press, 2009).

30 John Cassara, *Hide and Seek: Intelligence, Law Enforcement, and the Stalled War on Terrorist Finance* (Washington, DC: Potomac Books, 2006).

31 Timothy Wittig, *Understanding Terrorist Finance* (London: Palgrave MacMillan, 2011).

32 Nikos Passas, "Fighting Terror with Error: The Counter-productive Regulation of Informal Value Transfers," *Crime, Law and Social Change* 45, no. 4–5 (2006): 315–336.

33 Cassara, *Hide and Seek.*

34 For an extended discussion see Wittig, *Understanding Terrorist Finance,* chap. 2.

35 Frank M. Buchheit, "The Task Force Targeting Meeting: Operations Synchronization," *Infantry Magazine,* May–June 2008.

36 Wittig, *Understanding Terrorist Finance,* 2011, chap. 4.

EFFECT OF EXTERNAL ACTOR RESOURCES ON LOCAL ECONOMY

Guillermo Pinczuk

External aid can certainly play an important role in insurgencies, as it can allow both insurgent movements and governments to undertake actions they otherwise would not be able to carry out. In the case of both insurgents and counterinsurgents, external aid can be used to pay salaries of warfighters, underwrite a "hearts and minds" campaign, finance the purchase of necessary supplies, and help develop and disseminate propaganda targeted at specific audiences. Aside from these direct impacts, external aid can also have an indirect effect on a conflict, particularly with respect to decision making. For instance, DeVore noted that continued Iranian aid (and the provision of a safe haven in Iran) enabled Hizbollah to undertake long-range planning as well as promote organizational learning and the professionalization of personnel.[1]

External aid can also have an indirect economic impact. A classic example is the case of British aid to the Greek resistance against the Germans in World War II, where gold sovereigns used by the British to finance the resistance worsened an existing inflationary spiral.[2] This short paper looks at the issue of the impact, and especially the indirect impact, of external aid on a local economy. As a baseline, we first provide some information regarding Iranian financial assistance to Hizbollah to provide a glimpse into the importance of aid in modern insurgencies. Next we address several of the main unintended consequences of the provision of external aid. These include the infamous "Dutch disease," where external aid causes a reduction in a country's economic competitiveness; the impact of external aid on inflation and local spending; and the impact of short planning horizons on the efficient allocation of resources. Given the paucity of data on direct transfers to insurgent movements and their economic impact, this paper also looks at examples of external aid in the context of United Nations (UN) missions.

Although the fact that external aid can be of enormous assistance to an insurgent movement is fairly obvious, it is useful to look at the various dimensions of Iranian aid to Hizbollah to develop an appreciation for the scope of activities such aid enables. Levitt noted that Iran provides Hizbollah nearly $200 million per year, and this figure may be as high as $350 million.[3] In addition to funding salaries,[a] much of this aid is used by Hizbollah to fund various social services that have provided much-needed assistance to Lebanon's long-neglected Shia community. For instance, in addition to paying 100 percent of the medical

---

[a] A 1989 report noted that Hizbollah paid its militiamen $150–200/month, not including food and benefits.[4] DeVore noted that by offering this level of salary, the group was able to attract veteran Shiite fighters formerly employed by Amal and Palestinian groups.[5]

expenses of Hizbollah's injured fighters, Iran's Martyrs Foundation also pays 70 percent of the expenses for caring for injured civilians, and it built the al-Rasul al-Azam hospital in Dahiya, a southern suburb of Beirut, to handle these cases.[6] Harik also noted that the foundation paid monthly pensions as well.[7]

Additionally, when water and electricity services were cut in Dahiya in 1990 because of fighting, Iran helped finance the construction of four-thousand-liter water reservoirs in each district in Beirut's southern suburbs, and these were built by Reconstruction Campaign, a Lebanese charitable organization.[8] As of 2006, Reconstruction Campaign continued to be the source of drinking water for half a million people.[9] Furthermore, as of 2006, the Imam Khomeini Assistance Committee had granted more than 130,000 scholarships and provided interest-free loans, and the Martyrs Foundation, along with Reconstruction Campaign and Hizbollah, financed the construction of housing for families in need, particularly those whose houses were damaged in war.[10] Iranian financing is also used to fund al-Manar, Hizbollah's television station.[11]

Clearly, as the case of Iranian funding of Hizbollah shows, external financing can play an important role in funding an insurgent movement's operations and help it build legitimacy. Yet external aid can also generate various unexpected (and undesirable) higher-order effects. One such outcome is Dutch disease, an effect well known to economists who study the impact of foreign aid.[12] Dutch disease refers to the reduced competitiveness of various sectors of a country's economy after the receipt of external aid. More specifically, the inflow of aid can contribute to a rise in the real exchange rate, which in turn makes imports cheaper, thereby creating a disincentive for local entrepreneurs to make investments in those sectors that compete with them. This phenomenon was present in South Vietnam after the provision of US aid that accompanied the military presence during the Vietnam War,[13] and there is some indication that external aid has led to Dutch disease in Afghanistan after the toppling of the Taliban. Specifically, a March 2006 International Monetary Fund (IMF) report noted that Afghanistan's currency had appreciated by 20 percent in real terms in the thirty months that ended in September 2005, and the IMF attributed this rise to opium- and aid-related inflows.[14] This appreciation in the real exchange rate affected the external competitiveness of Afghanistan's economy.

Another potential impact, as the Greek example suggests, is inflation. First, it should be mentioned that spending enabled by external financing can potentially have a noticeable impact on a host economy's gross domestic product (GDP). Economists measure the impact of fiscal stimuli through what is known as the "Keynesian multiplier," which is an estimate of the number of times an additional dollar of fiscal stimulus cycles through an economy.[15] Using a multiplier of 1.5,[b] a March 2006 UN report noted that in four of nine missions (Kosovo, Burundi, Liberia, and East Timor), spending by the international mission boosted GDP by more than 6 percent.[17] Much of the spending was through mission subsistence allowances (MSAs), which represent spending by deployed UN staff. Specifically, MSAs constituted more than half of total spending in four missions and between 40 and 50 percent in another four missions.[18] The second-largest component was procurement, which contributed more than 40 percent of spending in four missions, followed by spending on national staff salaries.

Interestingly, mission spending did not have a noticeable economy-wide impact on inflation. Inflation rates actually decreased during UN missions in Haiti and the Democratic Republic of the Congo (DRC), and in the cases of Kosovo, Liberia, Sierra Leone, and the Ivory Coast, the inflation rate was below 15 percent during UN missions, and often well below that rate.[19] There is evidence, however, of localized inflation, particularly within housing sectors. For instance, the UN mission in Kosovo reported a doubling and tripling of rents during the period 2000–2002, but there is limited evidence that such localized inflation spread throughout the economy.[20]

One area where UN missions did have a noticeable impact was in the hiring of local staff. Although the amount spent on wages for local staff was much smaller than that spent on procurement or MSAs, UN missions typically offered wages that were much higher than those in the local public sector, which contributed to a brain drain. For instance, in 2005, UN wages for midlevel staff in the DRC were $763/month, whereas the comparable government wage was $25/month.[21] In Kosovo, in 2004, the UN wage was €790/month, and the comparable government wage for midlevel personnel was €145/month.[22] Anecdotal evidence indicated that these wage differentials led skilled labor to leave government jobs to join UN missions, in some cases accepting jobs for which they were overqualified.[23]

One case that may be relevant for special operators, particularly if they decide to fund rebel movements or incipient governments that establish civil administrations to fulfill government roles, is that of East Timor. In this mission, the UN hired a large local staff that drew significantly from the local supply of labor, and in the absence of a well-established local labor market, the UN mission played an active role in establishing a civil service pay scale for the East Timor Transitional Authority.[24]

---

[b] The authors of the UN report consulted with UN mission staff, the IMF, and World Bank field staff, who suggested a multiplier in the range of 1.5–2.0.[16]

Another aspect of UN missions that may be relevant for the special operator is that such missions typically are of short formal duration (usually twelve months, which can be extended), and this short time frame leads to a condensed planning cycle that may encourage economically inefficient investments. For instance, by 2001, East Timor's national electrical grid was destroyed, but UN managers were not permitted to make investments in fixed infrastructure, perhaps owing to fears over the sustainability of such investments once the UN had left.[25] Rather than build a power plant, the UN relied on large portable generators to provide power to UN facilities and important population centers. One result of this decision is that East Timor at the time had the most expensive power generation system in Asia, at $0.20/kwh (compared with $0.03 in China and $0.05 in Indonesia).[26]

In conclusion, while external financial aid can be an important component of an overall assistance effort that supports an insurgent movement or a government engaged in a counterinsurgent campaign, special operators will need to be mindful of the distortions and second-order effects such aid may have on a local economy.

NOTES

[1] Marc R. DeVore, "Exploring the Iran-Hezbollah Relationship: A Case Study of How State Sponsorship Affects Terrorist Group Decision-Making," *Perspectives on Terrorism* 6, no. 4–5 (October 2012): 85–107.

[2] D. M. Condit, *Case Study in Guerrilla War: Greece during World War II* (Washington, DC: Special Operations Research Office, American University, 1961), 9.

[3] Matthew Levitt, *Hezbollah: The Global Footprint of the Party of God* (Washington, DC: Georgetown University Press, 2013), 12.

[4] Judith Harik, "Hizbullah's Public and Social Services and Iran," in *Distant Relations: Iran and Lebanon in the Last 500 Years*, ed. H. E. Chehabi (London: I. B. Taurus & Co., 2006), 271.

[5] DeVore, "Exploring the Iran-Hezbollah Relationship," 95.

[6] Harik, "Hizbullah's Public and Social Services and Iran," 275.

[7] Ibid., 280.

[8] Ibid., 273.

[9] Ibid.

[10] Ibid., 282; and DeVore, "Exploring the Iran-Hezbollah Relationship," 94.

[11] Levitt, *Hezbollah: The Global Footprint of the Party of God*, 12.

[12] For a recent study looking into the impact of Dutch disease, see Raghuram G. Rajan and Arvind Subramanian, "Aid, Dutch Disease, and Manufacturing Growth," *Journal of Development Economics* 94, no. 1 (2011): 106–118.

[13] Ethan Kapstein and Kamna Kathuria, *Economic Assistance in Conflict Zones: Lessons from Afghanistan*, Center for Global Development Policy Paper 013 (October 2012), 4.

[14] *Islamic Republic of Afghanistan: 2005 Article IV Consultation and Sixth Review under the Staff Monitored Program—Staff Report; Staff Statement; Public Information Notice on the Executive*

Board Discussion; and Statement by the Executive Director for the Islamic Republic of Afghanistan, Country Report No. 06/113 (Washington, DC: International Monetary Fund, March 2006), 18.

[15] Michael Carnahan, William Durch, and Scott Gilmore, *Economic Impact of Peacekeeping* (United Nations, Department of Peacekeeping, March 2006), 15.

[16] Ibid.

[17] Ibid.

[18] Ibid., 16.

[19] Ibid., 11.

[20] Ibid., 12.

[21] Ibid., 32.

[22] Ibid.

[23] Ibid., 33.

[24] Ibid., 33–34.

[25] Ibid., 43.

[26] Ibid.

HOW RESISTANCE GROUPS GENERATE LEGITIMACY

W. Sam Lauber and Margaret McWeeney

Legitimacy in international relations is a controversial subject and often a source of debate in the international community. At the state level, the standard for legitimacy is developed through a history of territorial wars, judicial decisions, and international agreements. This history provides a relatively clear standard for the legitimacy of state government, but the legitimacy of a group, cause, or organization remains less clear. This lack of clarity is problematic because states' standing governments continue to face challenges from domestic resistance groups, or efforts organized by a country's civil population to resist the legally established government or an occupying power,[1] and to validate their challenge to power these groups often seek or rely on their legitimacy as perceived by domestic and international audiences. Establishing a clear definition or standard for the legitimacy of resistance groups will help those groups, as well as their proponents and opponents, understand how legitimacy is generated.

To a resistance group, legitimacy is a coveted position in which the group's claims are acknowledged, its enterprise is supported, and its causes are championed domestically, internationally, or both. This position enables the group to pose a stronger, more meaningful challenge to an incumbent government. This paper argues that legitimacy of resistance groups can come in two main forms: (1) others' recognition of the group as a powerful entity; and (2) validation and substantiation of the group's claim. The intent of this paper is to explore definitions of domestic and international legitimacy, discuss how legitimacy can be gained or lost, and identify methods that resistance groups often use to generate legitimacy with domestic and international audiences.

THE MEANING OF LEGITIMACY TO INTERNATIONAL AUDIENCES

Throughout much of the nineteenth and twentieth centuries, scholars and philosophers debated what constitutes legitimacy in international affairs. Robert Jackson argues that legitimacy is not a statement about fairness or justice, but it is about the degree of consensus between principal states. He states that legitimacy is ruled by the beliefs of the major powers and not by some universal right.[2] Martin Wight contributes that major global powers construct a mutually agreed upon international values system by which they can determine the rightness and fairness of actions. This accepted international values system includes a standard of civilization, adherence to rights norms, and good government. These prerequisites serve as the conditionality for international society to recognize a state's legitimacy.[3]

When major global powers agree on a working order and form an international community, they then determine the normative standards for legitimacy, including Wight's standard of civilization, adherence to rights norms, and good government. If a government or a group adheres to these standards, it is likely to achieve legitimacy in the eyes of the international audience, the community of states, which means its claim to power, its goals, and its means are validated. By defining the legitimacy of state governments, these scholars also help to advance a working definition for the concept of international legitimacy for non-state groups that challenge governments.

As mentioned, this paper contends that resistance groups can achieve two forms of legitimacy: first, in the form of others' recognition of the group as a relevant player, and second, in the form of substantive legitimacy. An examination of case studies illustrates that a resistance group gains recognition through engagement with the international audience. Governments outside the state in which the resistance takes place will either officially and formally recognize the resistance without necessarily taking a position on the validity of the group's aims, or they will engage with the group as necessary to accomplish objectives such as negotiating a cease-fire or coordinating humanitarian aid. The same case studies illustrate that a resistance group gains or loses substantive legitimacy in the form of support or condemnation, respectively, from major players in the international community, particularly Western powers or international institutions such as the United Nations (UN). Groups that conform to the normative standards set by the international community "pass the test" of legitimacy in the eyes of the global powers and thus gain their support. Importantly, resistance groups need only pass the test, not ace it.

Historical examples make clear that even when a resistance group has committed humanitarian atrocities, particularly during battle or wartime, if the group demonstrates an attempt to abide by international standards, norms, and institutions it can receive support from major players that then validate the group's aims. For instance, as will be explored in more detail later in the paper, the United States backed the resistance group RENAMO, or Resistência Nacional Moçambicana, in Mozambique, in the early 1990s in return for the group's acceptance of the Rome Peace Accords, reduction of its use of violence, and transformation into a political party. RENAMO's willingness to accept a peace agreement and encourage accountable and transparent governance in the region brought the group legitimacy in the eyes of the United States and other major players, who then helped support the group financially for more than a decade.[4]

Conversely, if a group does not adhere to these standards or refuses any appeals for reform, the international community will likely come to the consensus that the resistance group is illegitimate. This occurs, for instance, after accounts of a group's indiscriminate violence against civilian populations, suicide bombings, genocide, and use of child soldiers, even after major powers appeal to the group and demand that the group end those activities. Once the group is deemed illegitimate, the international community can take measures to oppose the group and to remove its territorial claim and leadership.

An example is the Taliban, a resistance group that formed to provide a governance solution to the chaos in southern Afghanistan after Russian occupation. Although it became the ruling authority in Afghanistan and was accepted as legitimate domestically, because of its brutal ruling tactics, human rights abuses, and harboring of terrorist organizations, the regime was dismantled by the US global War on Terror and subsequent occupation of Afghanistan in 2001.[5]

This example helps illustrate that when a group openly defies the standards of substantive legitimacy as defined by the international community, the group makes itself vulnerable to delegitimation, universal condemnation, and elimination. Although the Taliban had gained legitimacy by recognition as the ruling authority of Afghanistan, it ignored the international standards against acts of terror and the use of indiscriminate violence. Because it refused to adhere to these standards, the Taliban had little to no chance of achieving international approval and substantive legitimacy, and this lack of legitimacy led to a military campaign to eliminate the Taliban forces.

THE MEANING OF LEGITIMACY TO DOMESTIC AUDIENCES

More abstract and philosophical is defining legitimacy at the domestic, community, or group level. Many political theorists recognize the definition proposed by Max Weber, which states, "Rule is legitimate when its subjects believe it to be so."[6] Thomas Franck contributes that group legitimacy does not have a definitive beginning but is a practice similar to accepting the Greenwich Mean Time, a social fact that is meaningful only to members of the community who accept it and those who can testify to the existence of that particular community.[7]

These scholars identify that a group can achieve legitimacy by recognition by being perceived as a relevant entity both from within a community and by other communities and powers. To the domestic audience this recognition and acceptance can come from other ethnic and religious groups, the general civil population, nongovernmental groups, or the incumbent government.

Through examples, this paper will also explore how groups achieve substantive legitimacy at the domestic level. This is achieved when a large base of supporters or citizens recognize that the group's claims are valid and compelling, as in the case of the Eritrean People's Liberation Front (EPLF). Groups also achieve domestic legitimacy by having a territorial or ethnic claim, as in the case of Hamas; or by proving to citizens that they can provide a viable alternative to the incumbent government, as in the case of the Tamil Tigers, or a better governance structure than the incumbent government, as in the case of Hizbollah.

While considering these understandings of legitimacy, the remainder of this paper aims to confront the challenge by identifying five concrete methods that have demonstrated importance in generating legitimacy in the eyes of domestic and international audiences, regardless of the cultural context. The paper will conclude with a look at a contemporary resistance group and its struggle for legitimacy within the international community.

METHODS OF ACHIEVING LEGITIMACY IN THE EYES OF DOMESTIC AUDIENCES

After discussing how groups achieve or fail to achieve substantive legitimacy, this paper will identify five ways that resistance groups achieve legitimacy, by recognition and/or substantive validation of claims, in the eyes of domestic and international audiences, which often prove mutually reinforcing. Therefore, the next sections will include a definition of the method of achieving legitimacy, the method's contribution to a group's legitimacy by recognition and to its legitimacy as substantive validation of a claim, a description of the mechanism by which the group achieves the method, and an example of its use.

Creating a Cause through the Use of Narrative

The first method by which a resistance group can achieve legitimacy is by creating a cause for action. The more attractive this cause, the greater chance that the resistance group will be successful in appealing to the civil population over the incumbent, or standing, government. A cause affords a resistance group the psychological force to "pry the population away from the [incumbent], to control it, and to mobilize it."[8] "With a cause, the [resistance] has a formidable, if intangible, asset that [it] can progressively transform into concrete strength."[9]

An attractive and relatable cause creates legitimacy by recognition by producing a foundation, a purpose for the group's existence, and a rallying flag for the group's fight. This also contributes to substantive legitimacy by portraying to domestic and international audiences the reason behind the group's movement. It illustrates that the resistance group is not fighting for sport or for a frivolous cause but rather for an identifiable reason. This illustration allows the group to gain traction with civilians who identify with its cause.

An important mechanism in the establishment of a cause is creating a compelling narrative that attracts the civilian population to the movement. The narrative is a "story [that an] armed struggle uses to justify its actions in order to attain legitimacy and favor among relevant populations."[10] Many resistance groups create this narrative to develop a group identity and an "us" versus "them" mentality. Resistance groups might strengthen their narratives through the use of websites, flyers, movies, propaganda materials, protests, and memorialization of fallen supporters of the resistance group.[11]

The EPLF provides an example of establishing a cause by creating a narrative to appeal to a population. The EPLF's struggle for independence began in 1952, after the British relinquished the colonies of Ethiopia and Eritrea. Immediately after, the Ethiopian government took full control of Eritrea, banning the Eritrean flag and outlawing the Eritrean language. In response, a small group of Eritrean rebels, formed under the banner of the Eritrean Liberation Front (ELF), fired on Ethiopian police forces in 1961, inciting a thirty-year struggle for Eritrean independence.[12]

From the onset, the ELF-turned-EPLF created a nationalist narrative to challenge "Ethiopian colonialism" and "expansionist hegemony." This narrative was reinforced by local anecdotes, stories, and songs and became part of the collective consciousness of the Eritrean people. Catchphrases such as, "one people, one nation," "independence of decision," "state sovereignty," and "self-reliant development" became the calling cards of the movement. Members of the group also referred to themselves as "freedom fighters," rather than as members of a resistance or rebel army.[13] Furthermore, during the 1960s and 1970s the EPLF fought against rival units and factions, eventually negotiating a unified movement under the EPLF command. This unification allowed the civilian population to focus on one group as the champion of independence.

In 1991, the EPLF took control of the city of Asmara, formed a provisional government, and attained de facto independence from Ethiopia. Ethiopia conceded to Eritrea's military victory and agreed to hold a UN-sponsored national referendum on statehood for Eritrea. The

EPLF's successful campaign of uniting Eritrea under one narrative over a thirty-year period contributed to 98.2 percent of registered voters in the state voting for Eritrea's independence, prompting the UN to recognize Eritrea's independence in 1992.[14]

Soon after Eritrea's independence, the EPLF evolved into a political party, the People's Front for Democracy and Justice (or PFDJ). In 1992, the leadership of the PFDJ transitioned unchallenged into the offices of the government, with the head of PFDJ, Isaias Afewerki, transitioning to the office of the president, a position he currently maintains.[15]

From its onset, the EPLF championed the cause of freedom by creating a narrative of one people united to relieve Ethiopian oppression. This narrative produced legitimacy by recognition by creating a unified and powerful front; it created substantive legitimacy by showing the local population why Eritrea deserved its independence and rallying the Eritrean population around the group. The fact that the EPLF was able to construct a relatable and compelling cause from the beginning of the conflict proved a critical building block in the achievement of domestic and international legitimacy and contributed to convincing domestic and international audiences of Eritreans' right to self-determination.

Creating a Loyal Base of Support by Creating a de Facto Government

The second common method of achieving legitimacy is for resistance groups to create a loyal and willing a base of supporters. A loyal base of support generates legitimacy by recognition by defining a territorial base of action and securing willing fighters for and donors to the cause. This following shows that the incumbent government has failed in some sense to fully represent, serve, or appeal to the nation's population and that the resistance group could provide a more equitable and competent governance structure. This base also creates substantive legitimacy by illustrating that the group is a viable and capable option to the incumbent government.

Creating such a base is commonly achieved by establishing a de facto government, or government in practice, which portrays the resistance group as a competent and reliable organization that can meet the everyday needs of the local population. Resistance groups have set up schools, job programs, medical facilities, charities, recreational clubs, and welfare programs to become indispensable parts of their communities, in the process making many civilians indebted followers.[16]

One example of this is Hamas, an organization that has maintained de facto authority, or authority in practice, over the Gaza Strip since

2007 largely due to its grassroots strategy and support base. Hamas, or Harakat al-Muqawama al-Islamiyya, is a Sunni Islamist group in Palestine that organized in 1987 to become the ruling authority in Palestinian-occupied territories and to drive Israeli forces from Palestinian territories of the Gaza Strip and the West Bank.[17]

To accomplish its goals, Hamas built an extensive infrastructure of social service organizations funded by neighboring Arab nations, such as Egypt and Qatar.[18] This infrastructure included schools, universities, refugee shelters, orphanages, relief funds, medical clinics, technical institutions, sports clubs, religious institutions, and women's institutions. Although these programs initially were intended to provide aid to their underserved population, Hamas exploited this infrastructure by recruiting tens of thousands of Palestinians for employment,[19] funneling money through charities and prompting recruits to use their salaries to support military activism. The group also used this infrastructure to generate a community of volunteers to offer logistical help and shelter to militant operatives, including hideouts for suicide bombers.[20] Citizens in Gaza also have turned to Hamas, feeling overlooked by the international community, particularly the United States and Europe, who traditionally have been outspoken in their support for Israel over Palestine.[21]

Over time, the de facto government's provision of services and security has demonstrated its competence, generated a devoted base of followers, and allowed the group to gain both recognized and substantive legitimacy in the eyes of the domestic audiences in the Gaza Strip.

As a result, the 2006 elections resulted in success for Hamas in the national legislature, the Palestinian Legislative Council (PLC). Hamas received 44.45 percent of the popular vote, won a majority of electoral districts, and installed leader Ismail Haniyeh as prime minister of the PLC.[22] However, as a governing party, Hamas soon isolated Palestinians further from the international community by refusing to recognize Israel, which brought a series of economic sanctions on the country. These sanctions led to a brief military conflict between Fatah and Hamas in 2007 and concluded with a cease-fire and negotiations toward a national unified government with a power-sharing agreement, known as the Mecca Agreement, between Hamas and Fatah.

Despite the military conflict, the large base of support and demonstration of competence in the Gaza Strip empowered Hamas to be a serious negotiating party rather than a group that could be easily dissolved by Fatah, Israel, or the United States. For these reasons, the negotiation resulted in Hamas being granted authority over the Gaza Strip and veto power over the actions of the PLC.[23] In sum, the base of support created by the de facto government generated legitimacy by recognition through the group's control of the territory of the Gaza Strip as a base of operations and control of a base of soldiers willing to fight. Its substantive legitimacy was illustrated through the Palestinians' perception that the group could serve as a viable alternative ruling party, as evidenced in the 2006 elections.

At the time of writing, Hamas's claims to legitimacy are being challenged as Israel commenced air strikes in the Gaza Strip to protect its border cities and remove Hamas's base of support. In six weeks of fighting, more than two thousand Palestinians have been killed and thousands more have fled their homes as cease-fire and international peace negations are failing to quell the fighting.[24] Although Hamas achieved domestic legitimacy by establishing a de facto government to create a loyal domestic base of support, it is still being challenged to prove its substantive legitimacy to the international community.

Developing an Armed Force that Can Battle the Incumbent by Leveraging the Resources of the Civilian Population

Many resistance movements form from neglect or even persecution by the incumbent regime. To combat continued mistreatment, resistance groups often develop an armed force dedicated to their cause and to overthrowing the incumbent. In turn, these armed forces help establish a group's legitimacy in the eyes of domestic audiences by making territorial gains in the country, getting the incumbent's attention, and, again, providing a viable threat to the incumbent regime.

The creation of armed forces also contributes to substantive legitimacy by illustrating the population's passion, commitment, and force behind the movement. By participating in an armed struggle, innocent people are willing to sacrifice their lives or take the lives of others for this cause. The sharing of such an intense belief in the rightness of a cause demonstrates that the cause resonates deeply in the community, thereby validating it. Furthermore, forming a sophisticated, hierarchical, and regimented armed campaign again demonstrates the capabilities and discipline of the resistance.

One way to establish an armed force is to leverage the resources of a civilian base. Establishing a militia requires resources, and many resistance movements draw on the civilian population to supply most of their resources.[25]

One example is the Liberation Tigers of Tamil Eelam (LTTE), a resistance group that developed from the Tamil ethnic group in Sri Lanka, more commonly known as the Tamil Tigers. The Tamil Tigers organized when Sri Lanka gained its independence in 1948 and the ethnic Tamil population became the target of discrimination and violence by the majority ethnic group, the Sinhalese. Due to a prior relationship with the nation's colonizers, approximately three hundred thousand Tamils had their citizenship revoked and lost their jobs in the civil service. Tensions escalated in 1981 when a Sinhalese mob ransacked Tamil-run businesses, including a sacred library that stored ancient Tamil literature. The Tamil population responded with a number of protests that evolved into a radical fringe group that mandated full independence for the Tamils.[26]

By 1983, the conflict had escalated into a guerrilla war against the Sinhalese government in the Tamil-held eastern Jaffna region.[27] The Tigers fueled their struggle through illicit trade, terrorism, and donors from their sizable diaspora community located primarily in Western Europe, North America, and Asia. At the group's height, it was reported that more than one million ethnic Tamils across the world provided support for the effort in Sri Lanka.[28] The group also extorted more moderate members of the Tamil ethnic group to raise money.[29] As a result, the Tamil Tigers generated between $200 million and $350 million per year (in 2008) to purchase arms and commit suicide bombings.[30]

The Tamils even leveraged the human capital of their community by recruiting or seizing Tamil children from a young age so that they could inculcate their nationalistic values and train them as soldiers/suicide bombers.

> Garlanded photos of young martyrs decorate the walls of Tamil Tiger training camps and Tamil-controlled cities and are regularly featured in Tamil newspapers . . . Another inducement is that bombers are granted the privilege of eating their last meal with the group's charismatic leader, Prabhakaran . . . Parents of the young martyrs are sometimes rewarded materially or are intimidated into relinquishing control over their children.[31]

After twenty years of fighting, the Sri Lankan government finally agreed to concessions making Tamil an official language of Sri Lanka, opening roads and air traffic to the Tamil-held Jaffna region, and allowing Tamils to serve in the civil service.[32] The Tamils and the incumbent government also agreed to a federal solution, which ended the Tamils' claim to an independent state.

However, sporadic fighting continued until 2009, when the Tigers' rebel commander Karuna was sworn in as minister of national integration and the Tigers were defeated in the last rebel-held territory in the northeast, resulting in the death of Tamil leader Prabhakaran. Since that time, the Tigers have transformed into a political entity, known as the Tamil National Alliance, which governs two-thirds of local councils in the north and east of the country.[33]

Despite their inability to realize their goal of an independent state, the Tamil Tigers leveraged the civilian population to develop a capable and competent armed force and generate domestic legitimacy, through both domestic recognition as a powerful player and substantive validation of their claim. This legitimacy was proven by the concessions the incumbent government made to the group. Although at times coerced, the local and diaspora populations formed the base of a significant military movement, which held territory in the eastern region, gained attention as a formidable movement from the Sri Lankan government, showed the group's commitment to its cause, and posed a viable threat to the Sri Lankan government.

METHODS OF ACHIEVING INTERNATIONAL LEGITIMACY

In addition to methods that contribute to legitimacy in the eyes of the domestic audience, this paper also describes two methods that resistance groups use to achieve legitimacy in the eyes of international audiences.

Developing a Capable Nonviolent Arm that Can Resist, Disrupt, or Challenge the Incumbent and Its Policies by Transforming into a Political Entity

A resistance group can achieve international legitimacy by developing a nonviolent campaign that demonstrates that the resistance movement has a moral, ethical, or intellectual rationale for its grievance and capable, rational leadership willing to work toward a negotiated resolution in a fairly contested system of government.

This nonviolent campaign not only works to bring recognition to the group through actions such as political rallies, protests, and participation in elections, but it also brings substantive legitimacy to the group's claim. A nonviolent campaign illustrates that the resistance group is not fighting solely for the sake of power and greed but instead is fighting for a moral or ethical struggle; it also illustrates that the group has rational leadership looking to seek some form of resolution from the incumbent.

A group can achieve legitimacy by transforming itself from an armed resistance group into a political party. This evolution into a political party may occur only after substantial international pressure, but it demonstrates that a violent campaign waged during wartime is an unfortunate consequence of being marginalized or treated poorly by the incumbent government. Second, it compels the resistance movement to gain the favor of the international community by identifying, recruiting, and retaining competent candidates for public office.[34] Finally, it conveys to the international community and domestic audience that if a state is willing to transform into a more equal and democratic system of governance, the resistance group is willing to work toward a political and negotiated settlement rather than a violent solution to its grievance.[35]

One group that consented to international pressure to evolve into a nonviolent campaign is RENAMO. RENAMO is an anti-communist group that used violence to gain traction over the leading party, Front for the Liberation of Mozambique, or FRELIMO, after Mozambique's declaration of independence in 1975. As a consequence of FRELIMO's tight control of the governing apparatus, a civil war developed between the two groups in which both FRELIMO and RENAMO were supported by opposing international powers, including the United States, Great Britain, the Soviet Union, and South Africa.[36]

After a decade of attacks on Mozambique's infrastructure and eight rounds of peace talks, the UN facilitated the Rome General Peace Accords in 1992 where RENAMO was persuaded to cease guerrilla warfare and embrace political organization to pursue its claim to power.[37] As part of the agreement the party received a $17 million trust fund from multiple international donors and $55 million to defray the cost of elections.[38] International sponsors of the peace process accepted the core leadership of RENAMO, including party President Afonso Dhlakama,[39] although the group was compelled to use the trust fund to replace most of its internal organization with legitimate administrators and legislative candidates and encourage a commitment to democratic system of government.[40] As a result, in Mozambique's first democratic elections, RENAMO was successful in gaining 112 of 250 legislative representatives,[41] and by the second election in 1999, RENAMO had cemented its position in the two-party system in Mozambique.[42]

Although RENAMO initially had support from the United States, after twenty years of fighting, the group faced the decision either to reform or to risk termination and international condemnation for its human rights abuses. By accepting the tenets of nonviolence in the Rome Accord and transforming into a political entity, RENAMO garnered international legitimacy as a recognized player and substantive legitimacy by being willing to work toward a peaceful solution to its claim to power. It achieved this legitimacy by developing an intellectual party platform, replacing its internal organization with competent leadership, and participating in free and fair elections. Legitimacy has allowed the group to receive ample external funding to remain a politically salient resistance movement for more than thirty years.

Maintaining Relationships with States and International Political Institutions by Securing Support or Sponsorship through Moral, Political, Technical, Financial, and Military Means

Resistance groups can also generate and maintain legitimacy in the eyes of the international audience by maintaining relationships with states and international political institutions. These relationships allow the group legitimacy by recognition by propelling it to the international stage as a movement that spans regions or countries, while financial, technical, and military support can sustain the group's presence for long periods, increasing its chances of achieving recognition.

This method also contributes to substantive legitimacy by demonstrating to the domestic audience and other international forces that the group has significant financial backing. This financial support suggests that there is a stake in the group's success from major players in the international community and that the group could present a sustainable and credible alternative to the incumbent.

Many resistance groups maintain relationships with states and international institutions by securing sponsorship from an external donor or a variety of donors. In exchange for this support, resistance groups often act as a strategic arm of these donors, who may be reluctant to challenge another government on the international stage or in a traditional territorial war.

An example of this can be seen in the actions of Hizbollah, or "Party of God," a Shia Islamic resistance group in Lebanon. The group came into prominence in the early 1980s when Israel invaded and occupied Lebanon to expel Palestinian militants operating in the south. In response to the occupation, Hizbollah began a violent campaign against the Israeli military.

Since then, Hizbollah has become known as the strategic arm for Iran in the Arab world, receiving financial support and training from Iran's Revolutionary Guard. As part of the trade-off for such support, the organization vows loyalty to Iran's Ayatollah Khomeini and aims to expel the influence of Western powers from the Arab world.[43]

Annual estimates of funding from Iran vary from tens of millions of dollars to one billion dollars, not including military training and sophisticated weapons systems. Other contributors to Hizbollah include wealthy Arab individuals, such as Mohammed Hammoud who was convicted in 2003 for accumulating millions of dollars in profit for Hizbollah, as well as a sizeable Lebanese expatriate community, which has been known to funnel illicit funds from the South American drug trade into Lebanon.[44]

Over thirty years, Hizbollah has achieved de facto control of southern Lebanon, leveraging its massive resources to become an indispensable part of the Lebanese community in areas underserved by the current government in south Lebanon and Beirut,[45] and has transformed into a competent political and military entity.[46] In 2006, Hizbollah fended off Israeli forces from southern Lebanon, and in 2009, Hizbollah won ten parliamentary seats in the Lebanese parliamentary elections under leader Hassan Nasrallah.[47] By securing external support from Arab donors and states through political, technical, financial, and military means, Hizbollah has been propelled to the international stage as a leading player in the Arab world and has exhibited staying power for more than thirty years despite international and domestic attempts to dismantle the group. Additionally, Hizbollah's close ties and sponsorship from Iran, specifically, show the incumbent government and international community that it has substantial weight and force behind its movement and that it will not be easily eliminated. These resources, and the political support of countries such as Iran, forced the incumbent government and international community to recognize Hizbollah's formidable strength and resulted in the domestic audience's increased approval of the movement as a credible alternative.

CONCLUSION

By defining legitimacy in two ways, (1) recognition and (2) substantiation of claim, this paper discusses ways in which a resistance group gains or loses legitimacy and describes five common methods of achieving legitimacy in the eyes of domestic and international audiences. However, this paper aims only to contribute to the debate on the meaning or attainment of legitimacy by resistance groups.

Legitimacy remains an ambiguous concept, as evidenced by the recent events in Ukraine. In February 2014, the Ukrainian Parliament deposed its president and appointed an interim government under Oleksandr Turchynov, a leader recognized by the United States and European Union. However, Russia condemned this move and retaliated by taking control of the Crimean peninsula, a Ukrainian territory with an ethnic Russian majority. The Crimean Parliament voted to dismiss its government and call a referendum on Crimean independence. This referendum was vastly in favor of declaring independence from

Ukraine but was subsequently condemned by the European Union, the United States, and Ukraine.[48]

In March 2014, the separatist government of Crimea signed a treaty of accession into the Russian Federation, although the UN General Assembly declared the treaty invalid and incorporation by Russia illegal.[49] The United States and European Union enacted sanctions against Russia,[50] although a handful of countries, including Afghanistan, Venezuela, and Syria,[51] have come out in support of the Crimean referendum. Russia has also heavily armed the area and is accused of inciting and orchestrating the revolt. This crisis has brought even greater attention as Crimean rebels have responded with acts of terrorism including downing a Ukrainian plane, a military helicopter, and a Malaysian airliner carrying almost three hundred civilian passengers.[52]

In sum, there remains no formula for achieving legitimacy. Application of these five methods and adherence to international rights norms will not make a resistance group a relevant domestic player that has just cause for its grievance overnight. However, this paper aims to contribute to the debate on what constitutes legitimacy by discussing paths to substantive legitimacy and identifying five common methods and mechanisms that have increased the likelihood of achieving domestic and international legitimacy for a number of resistance movements.

Table 1. Five ways resistance groups achieve legitimacy.

Group	Intended Audience	Method	Rationale for Legitimacy	Mechanism
EPLF	Domestic	Establishing a relatable and compelling cause	Creating a cause creates a purpose for the group's existence and a rallying flag for its fight and generates a base of support.	Creating an attractive narrative
Hamas	Domestic	Generating a loyal base of support	Generating a base of support allows the group a territorial base, secures fighters and donors, and depicts the group as an alternative to the standing government.	Establishing a de facto government
LTTE	Domestic	Developing a capable and competent armed force to battle the incumbent	Developing an armed force allows the group to gain or maintain territory, grab the government's attention, convey itself as a viable threat, and demonstrate the passion, commitment, and discipline behind the movement.	Using and leveraging the resources of the civilian population (including diaspora population)

Group	Intended Audience	Method	Rationale for Legitimacy	Mechanism
RENAMO	Domestic/ international	Developing a capable non-violent arm that can resist, disrupt, or challenge the incumbent	Developing a nonviolent arm conveys that the group maintains an intellectual, moral, or ethical arm and that the group has capable and competent leadership willing to be reasonable and to participate democratically in a fairly contested system of government.	Transforming into a political party
Hizbollah	International	Maintaining relationships with states and international political institutions	By maintaining relationships with or sponsorship by external donors and states, the group is propelled to the international stage while financial, technical, and military support can sustain the group's presence for long periods.	Securing external support through political, technical, financial, and military means

Source: Author's synthesis and "Guide to the Analysis of an Insurgency." *United States Marine Corps* (2012).

NOTES

[1] Joint Publication 1-02 (JP 1-02), *Department of Defense Dictionary of Military and Associated Terms* (Washington, DC: Department of Defense, November 8, 2010, as amended through August 15, 2014), http://www.dtic.mil/doctrine/new_pubs/jp1_02.pdf.

[2] Robert H. Jackson, *The Global Covenant: Human Conduct in a World of States* (Oxford: Oxford University Press, 2000), 91.

[3] Martin Wight, "International Legitimacy," *International Relations* 4, no. 1 (1972): 27.

[4] Carrie Manning, "Armed Opposition Groups into Political Parties: Comparing Bosnia, Kosovo, and Mozambique," *Studies in Comparative International Development* 39, no. 1 (2004): 54–76.

[5] Zachary Laub, "Backgrounder: The Taliban in Afghanistan," Council on Foreign Relations, updated July 4, 2014, http://www.cfr.org/afghanistan/taliban-afghanistan/p10551.

[6] Max Weber, *On Charisma and Institution Building* (Chicago: University of Chicago Press, 1968), 11.

[7] Thomas Franck, *The Power of Legitimacy Among Nations* (New York: Oxford University Press, 1990), 204.

[8] David Galula, *Counterinsurgency Warfare: Theory and Practice* (Westport, CT: Praeger Security International, 2006), 11.

[9] Ibid.

[10] US Government, *Guide to the Analysis of an Insurgency 2012*, 6, http://www.mccdc.marines.mil/Portals/172/Docs/SWCIWID/COIN/Doctrine/Guide%20to%20the%20Analysis%20of%20Counterinsurgency.pdf.

[11] Ibid.

[12] Charles E. Cobb Jr. and Roberto Caputo, "Eritrea Wins the Peace," *National Geographic* 189, no. 6 (1996): 82–106.

[13] Andebrhan Welde Giorgis, *Eritrea at a Crossroads: A Narrative of Triumph, Betrayal, and Hope* (Houston: Strategic Book Publishing and Rights, 2014), 401.

[14] Ibid., 161.

[15] "World Report 2013: Eritrea," Human Rights Watch, January 1, 2013, http://www.hrw. org/world-report/2013/country-chapters/eritrea.

[16] Ibid., 2.

[17] "Who are Hamas?" *BBC News*, updated January 4, 2009, http://news.bbc.co.uk/2/ hi/1654510.stm.

[18] Matthew Levitt and Dennis Ross, *Hamas: Politics, Charity, and Terrorism in the Service of Jihad* (The Washington Near East Institute, 2006), 26–28; and Jonathan Masters and Zachary Laub, "Hezbollah (a.k.a. Hizbollah, Hizbu'llah)," Council on Foreign Relations, January 3, 2014, http://www.cfr.org/lebanon/hezbollah-k-hizbollah-hizbullah/p9155.

[19] Scott Wilson, "In Politics, Hamas Gains in the West Bank," *Washington Post*, June 29, 2005, http://www.washingtonpost.com/wp-dyn/content/article/2005/06/28/ AR2005062801368.html.

[20] Levitt and Ross, *Hamas: Politics, Charity, and Terrorism in the Service of Jihad*, 51.

[21] "Middle East: Why Did Hamas Win Palestinian Poll?," Radio Free Europe/Radio Liberty, January 26, 2006, http://www.rferl.org/content/article/1065113.html.

[22] Scott Wilson, "In Politics, Hamas Gains in the West Bank."

[23] "Who are Hamas?" *BBC News*; and Jim Zanotti, "Hamas: Background and Issues for Congress," CRS Report no. R41514 (Washington, DC: Congressional Research Service, December 2, 2010), http://www.fas.org/sgp/crs/mideast/R41514.pdf, 3–4.

[24] Katherine Butler, "Why Did Israel Attack Gaza?" *Huffington Post*, posted January 29, 2009 and updated May 25, 2011, http://www.huffingtonpost.com/2008/12/29/why-did-israel-attack-gaz_n_153987.html; and Karin Laud and Josef Federman, "Israel Calls for North Gaza Evacuation after Raid," *Huffington Post*, posted July 13, 2014 and updated September 11, 2014, http://www.huffingtonpost.com/2014/07/13/israel-gaza-evacuation-raid_n_5581540.html.

[25] US Government, *Guide to the Analysis of an Insurgency*, 11.

[26] Sumana Raychaudhuri, "Will Sri Lanka Drive the Tigers to Extinction?" *The Nation*, February 6, 2009, http://www.thenation.com/article/will-sri-lanka-drive-tigers-extinction.

[27] Ibid.

[28] Ibid.

[29] Randall Law, *Terrorism: A History* (Cambridge, UK: Polity Press, 2009).

[30] Paul Collier, Anke Hoeffler, and Dominic Rohner, "Beyond Greed and Grievance: Feasibility and Civil War," *Oxford Economic Papers* 61, no. 1 (2009): 4.

[31] Law, *Terrorism: A History*.

[32] Ibid.

[33] Ibid.

[34] Manning, "Armed Opposition Groups into Political Parties," 59.

[35] Ibid., 57.

[36] James M. Scott, *Deciding to Intervene: The Reagan Doctrine and American Foreign Policy* (Durham, NC: Duke University Press, 1996), 194–195 and 197.

[37] "Terrorist Organization Profile: Mozambique National Resistance Movement," National Consortium for the Study of Terrorism and Responses to Terrorism, http://www.start. umd.edu/tops/terrorist_organization_profile.asp?id=314.

[38] Christoph Zürcher et al., *Costly Democracy: Peacebuilding and Democratization after War* (Stanford: Stanford University Press, 2013), 104.

[39] Manning, "Armed Opposition Groups into Political Parties."

[40] Carrie Manning, *The Making of Democrats: Elections and Party Development in Postwar Bosnia, El Salvador, and Mozambique* (New York: Palgrave Macmillan, 2008), 48–53.

[41] "Terrorist Organization Profile: Mozambique National Resistance Movement," National Consortium for the Study of Terrorism and Responses to Terrorism; Gwinyayi Dzinesa and Paulo Wache, "Mozambican Elections: What to Make of Dhlakama's Intention to Run for President," *All Africa*, May 27, 2014, http://allafrica.com/stories/201405270478.html; and Manning, "Armed Opposition Groups into Political Parties."

[42] Manning, *The Making of Democrats*, 56.

[43] Masters and Laub, "Hezbollah (a.k.a. Hizbollah, Hizbu'llah)."

[44] US Department of Homeland Security, Immigration and Customs Enforcement, "Mohamad Youssef Hammoud Sentenced to 30 years in Terrorism Financing Case," news release, January 27, 2011, http://www.ice.gov/news/releases/1101/110127charlotte.htm.

[45] "Lebanon: The Many Hands and Faces of Hezbollah," *IRIN*, March 29, 2006, http://www.irinnews.org/report/26242/lebanon-the-many-hands-and-faces-of-hezbollah.

[46] Iain Scobbie, "South Lebanon 2006," in *International Law and the Classification of Conflicts*, ed. Elizabeth Wilmshurst (Oxford: Oxford University Press, 2012), 403.

[47] Masters and Laub, "Backgrounder: Hezbollah"; and Shana Marshall, "Hizbollah: 1982-2009," in *Casebook on Insurgent and Revolutionary Warfare, Volume II: 1962–2009*, ed. Chuck Crossett (Fort Bragg, NC: United States Army Special Operations Command, 2012), 560.

[48] Alissa de Carbonnel, "RPT-INSIGHT—How the Separatists Delivered Crimea to Moscow," *Reuters*, March 13, 2014, http://in.reuters.com/article/2014/03/13/ukraine-crisis-russia-aksyonov-idINL6N0M93AH20140313; and "Ukraine Crisis: Russia Isolated in UN Crimea Vote," *BBC News Europe*, March 15, 2014, http://www.bbc.com/news/world-europe-26595776.

[49] Somini Sengupta, "Vote by U.N. General Assembly Isolates Russia," *New York Times*, March 27, 2014, http://www.nytimes.com/2014/03/28/world/europe/General-Assembly-Vote-on-Crimea.html.

[50] Kimberly Marten, "Why Sanctions against Russia Might Backfire," *Huffington Post*, August 21, 2014, http://www.huffingtonpost.com/kimberly-marten/why-sanctions-against-rus_b_5696038.html.

[51] Matthew Rosenberg, "Breaking with the West, Afghan Leader Supports Russia's Annexation of Crimea," *New York Times*, March 23, 2014, http://www.nytimes.com/2014/03/24/world/asia/breaking-with-the-west-afghan-leader-supports-russias-annexation-of-crimea.html?ref=asia&_r=0.

[52] Richard Balmforth and Pavel Polityuk, "Ukraine: Rebels Have Shot down a Ukrainian Military Plane," *Huffington Post*, August 7, 2014, http://www.huffingtonpost.com/2014/08/07/rebels-ukraine-plane_n_5659861.html.

TRANSITION FROM RESISTANCE TO GOVERNANCE

W. Sam Lauber and Margaret McWeeney

INTRODUCTION

Contemporary scholarship on resistance movements has yet to comprehensively explain methods by which resistance groups successfully transition to governing authorities. Examples such as the American Revolution, in which a resistance group transitioned into a sustainable governance structure, have proven rare throughout history. In fact, most literature on resistance movements focuses on the group's development in stages of governance, such as attainment of de facto authority (authority in practice) in a region, achievement of political representation in a national legislature, and finally, establishment of a new governing authority or creation of an independent state.

Through a careful analysis of seven contemporary resistance groups, this paper distills three commonalities that are found to be major contributing factors in a resistance group's transition through these stages of governance: (1) a base of support for the resistance group; (2) the group's relationship with internal and external donors; and (3) an internal organization and administrative structure.

BASE OF SUPPORT

The first factor contributing to transition from resistance to governance is a base of support, whether local, regional, or national. Resistance groups must form a base of support to bolster their claims, recruit volunteers, and support their movements. To achieve this support, resistance groups often exploit weaknesses in the incumbent regime, typically by providing goods or services that the incumbent regime fails to provide or that the resistance group feels are not being offered to the local population.[1]

Hamas

One illustrative case is Hamas, an organization that has maintained de facto authority of the Gaza Strip since 2007 due to its grassroots strategy and support base. Hamas, or Harakat al-Muqawama al-Islamiyya, is a Sunni Islamist group in Palestine that organized in 1987, at the beginning of a Palestinian uprising against the Israeli state. Its goals are to drive Israeli forces from Palestinian territories of the Gaza Strip and the West Bank and to replace Fatah, the ruling party in Palestine.[2]

To accomplish its goals, Hamas began building an extensive infrastructure of social service organizations funded by neighboring Arab nations, such as Egypt and Qatar.[3] This infrastructure included schools, universities, refugee shelters, orphanages, relief funds, medical clinics, technical institutions, sports clubs, religious institutions, and women's institutions. Hamas also recruited tens of thousands of Palestinians for employment,[4] began funneling money through its charities, and prompted recruits to use their salaries to support military activism. Furthermore, Hamas used this infrastructure to generate a community of volunteers to offer logistical help, shelter, and hideouts to militant operatives such as suicide bombers.[5] Over time, the provision of services allowed Hamas to gain the loyalty of citizens in the Gaza Strip who had grown weary of dealing with the corrupt and neglectful ruling party, Fatah. Citizens in Gaza also turned to Hamas as the international community, particularly the United States and Europe, had become outspoken in its support for Israel.[6]

As a result, the 2006 elections resulted in success for Hamas in the national legislature, the Palestinian Legislative Council (PLC). Hamas received 44.45 percent of the popular vote, won a majority of electoral districts, and installed leader Ismail Haniyeh as prime minister of the PLC.[7] This was a shock to both the international community as well as the incumbent government.

As a governing party, Hamas soon isolated Palestine further from the international community by refusing to recognize Israel, which brought a series of economic sanctions on the country. Incurring these economic sanctions led to a brief military conflict between Fatah and Hamas in 2007 that concluded with a cease-fire and internationally facilitated negotiations toward a national unified government with power sharing between Hamas and Fatah, known as the Mecca Agreement.

From its inception, Hamas worked diligently to become an indispensable player in Palestine and garner the loyalty of Palestinian citizens. Hamas exploited the weakness of the ruling party, Fatah, providing social services and employment opportunities to underserved areas. By doing so, Hamas created a domestic base of support that voted favorably for the organization in the 2006 elections. Hamas was able to resist Fatah's military efforts to dismantle the group and was recognized by the international community as a relevant domestic player. Because of Hamas's significant base of support in the region, it was brought to the table in international negotiations and offered a power-sharing agreement, giving it authority over the Gaza Strip and veto power over the actions of the PLC.[8]

The Taliban

A second example is the Taliban, a resistance group that came into prominence by exposing the corruption and misrule of the incumbent government in Afghanistan. The Taliban is also a good example of a resistance group that was able to gain support of the local population, despite its extreme ideology, by providing security to local civilian populations in war-torn regions.

The Taliban's strong hold in Afghanistan began after the Soviet Union withdrew from the country in 1989 and its communist, Soviet-led successor government fell in 1992. When the Soviet government collapsed, several mujahideen, or resistance fighting groups, signed the Peshawar Accord to appoint a national government for the Islamic State of Afghanistan under interim President Burhanuddin Rabbani, a Tajik mujahideen leader.[9]

This government soon collapsed, plunging the country into chaotic civil war. During the civil war, the country was divided into "warlord fiefdoms . . . and all the warlords fought, switched sides and fought again in a bewildering array of alliances, betrayals, and bloodshed."[10] Southern Afghanistan and Qandahar were divided among dozens of ex-mujahideen warlords and bandits who plundered the population and prohibited international relief.

The Taliban formed from refugees and students in Pakistani madrassas to provide a solution to the chaos in southern Afghanistan. As a conglomeration of several groups of Pashtun origin, the Taliban aimed to fight against rampant corruption in the country, stop the conflict between the mujahideen warlords, and make Afghanistan an Islamic state.[11] Over time, the Taliban demonstrated its capacity to govern in southern Afghanistan, particularly by providing security to Afghan citizens. Civilians began to favor the Taliban and saw it capable of purging the mujahideen resistance movement, which had been corrupted with the power of governance. For instance, a mujahideen military commander was reported to have abducted and raped two teenage girls. In response, Taliban leader Mullah Omar enlisted a small militia to attack and hang the commander and free the girls. From actions like this, Taliban members gained the mystique of "Robin Hood figures."[12]

Soon after, the Taliban was seen as the ruling authority in southern Afghanistan. In 1994, Pakistani Prime Minister Benazir Bhutto enlisted the Taliban's help, rather than that of the Afghan government, to secure a trade route from Peshawar to Kabul. Pakistani trucking companies donated several hundred thousand Pakistani rupees (millions of US dollars) and promised a monthly stipend to the Taliban to clear the roads and guarantee security. Additionally, the Pakistani government gave the Taliban tens of thousands of pieces of artillery.[13]

The Taliban formally secured Qandahar in 1994 backed by Pakistan's intelligence service and endowments from Saudi Arabia. In doing so, it captured the city's tanks, armored cars, military vehicles, weapons, and helicopters. Leveraging power from controlled territories in the south, the Taliban was eventually successful in removing Afghan President Rabbani in 1996 and establishing the Islamic Emirate of Afghanistan.[14]

Due to the Taliban's brutal ruling tactics, human rights abuses, and harboring of terrorist organizations, the regime was dismantled during the US occupation of Afghanistan in 2001, sending many former Taliban leaders into neighboring Pakistan.[15]

The Taliban is a clear example of a resistance group that achieved control and eventually became the state authority by providing goods and services and aligning with the local population. Even if Afghan individuals did not agree with or approve of the Taliban's ruling tactics or ideology, communities were compelled to support the group because of the egregious neglect and corruption of the standing government. Even after the US occupation of Afghanistan and the declared global War on Terror, the Taliban still remains a prominent movement in northern Pakistan. The Taliban example provides evidence that assuming the role of government for underserved populations can generate a domestic base of support that can propel a resistance group into a governing authority.[16]

RELATIONSHIP TO DONORS

To provide goods and services and gain de facto authority, resistance groups must finance their operations. Therefore, the second major factor contributing to a resistance group's ability to gain governing authority is its relationship with internal and external donors.[17] Extensive funding enables resistance groups to overcome various obstacles and opposition, as well as to try alternative avenues to stay in power.

Hizbollah

One example is Hizbollah, or "Party of God," a Shia Islamic resistance group in Lebanon. The group came into prominence in the early 1980s when Israel invaded and occupied Lebanon to expel Palestinian militants operating in the south. In response to the occupation, the Hizbollah movement began a violent campaign against the Israeli military.

Soon after, Hizbollah became the strategic arm for Iran in the Arab world, receiving financial support and training from Iran's Revolutionary Guard. In fact, in the party's manifesto, the organization vows loyalty to Iran's Ayatollah Khomeini and aims to expel the influence of the United States, France, and Israel from Lebanon.[18]

Annual estimates of funding from Iran vary from tens of millions of dollars to one billion dollars, not including military training and sophisticated weapons systems. Other contributors to Hizbollah include wealthy Arab individuals, such as Mohamad Hammoud, who was convicted in 2003 of accumulating millions of dollars in profit for Hizbollah, as well as a sizeable Lebanese expatriate community, which has also been known to funnel illicit funds from the South American drug trade into Lebanon.[19] Hizbollah also works to recruit Lebanese professionals from government institutions, nongovernmental organizations, and universities and encourage them to contribute. This money, manpower, and artillery fuels Hizbollah's organizational capacity to gain a presence and maintain de facto control in southern Lebanon, gain the trust of the Lebanese population, and propel its political agenda.[20]

However, over the group's thirty-year development, Hizbollah leaders have come to understand that to survive the group must adopt legitimate practices, particularly because Israel is backed by the United States and because Lebanese citizens prefer not to associate with a terrorist organization.[21] To address this, Hizbollah leveraged its massive resources to establish itself as a legitimate resistance movement instead of a terrorist group. Hizbollah developed an extensive social development program with four hospitals, twelve clinics, twelve schools, and two agricultural centers for assistance and training. The organization targeted these programs to address areas underserved by the current government in southern Lebanon and Beirut.[22]

Hizbollah's demonstration of its ability to operate as an alternative to the ruling regime, combined with a 2006 military conflict in which it fended off Israeli forces in southern Lebanon, increased its popularity in the country and won the group ten parliamentary seats in the 2009 Lebanese parliamentary elections under leader Hassan Nasrallah.[23] Although still a minority party, Hizbollah has been successful in leveraging its resources from donors to become a de facto government in southern Lebanon and gain political representation in the national legislature. The surplus of resources from Iran, individual donors, and the Lebanese expatriate community continue to allow Hizbollah to rule southern Lebanon and operate as a "state-within-a-state,"[24] despite international pressures to dismantle the group because of its anti-Western agenda.

RENAMO

For the vast majority of conflicts that end in peace negotiations rather than military victory, negotiators generally aim to entice aggressive resistance groups to end their violent tactics in exchange for becoming legitimate political parties. Often external actors, such as international organizations like the United Nations (UN) or states with a particular interest, will offer financial incentives to encourage resistance groups to take that deal.

One group that accepted such an agreement is RENAMO, or Resistência Nacional Moçambicana, in Mozambique. RENAMO is an anti-communist organization that historically used violence to gain traction over the leading party, Front for the Liberation of Mozambique, or FRELIMO. FRELIMO became the ruling authority after Mozambique's declaration of independence in 1975 and maintained tight control of the governing apparatus. As a consequence, a civil war developed between the two parties in which both FRELIMO and RENAMO became pawns of international players, including the United States, Great Britain, the Soviet Union, and South Africa.[25]

For instance, the Soviet Union entered into a "Treaty of Friendship" with the socialist FRELIMO, and Great Britain became sympathetic to the claims of FRELIMO after witnessing the kidnapping of British nationals abroad by rebel groups. RENAMO received financial and military support from conservatives within the Reagan White House in the United States and investment from Rhodesia and South Africa to punish FRELIMO for supporting national liberation movements such as the African National Congress, the black anti-apartheid group in South Africa.[26]

After a decade of attacks on Mozambique's communication and trade infrastructure and eight rounds of peace talks, the UN facilitated the Rome General Peace Accords in 1992, in which RENAMO was persuaded to cease guerrilla warfare and embrace political organization to pursue its claim to power. RENAMO agreed to a cease-fire but only after it was guaranteed total amnesty for any violence that occurred during the civil war.[27]

As part of the agreement, the party received a $17 million trust fund from multiple international donors. Donors additionally paid $55 million to defray the cost of elections.[28] International sponsors of the peace process accepted the core leadership of RENAMO, particularly party President Afonso Dhlakama, and allowed leaders to maintain control over financial resources, allocation of positions, relations with external actors, and communication.[29] However, RENAMO was obligated to use the trust fund to replace most of its internal organization with legitimate administrators and legislative candidates and to encourage a commitment to a democratic system of government.[30]

In 1994, Mozambique held its first democratic elections. Despite FRELIMO's victory in the legislature, RENAMO was successful in gaining 112 of 250 legislative representatives/seats, and only 18 of these 112 representatives had participated in RENAMO during the civil war. [31] The majority of these positions were given to new recruits used to fill out candidate election ballots. By the second election in 1999, RENAMO's representatives had gained visibility and professional experience within both the country and the international community and had cemented the two-party system in Mozambique. RENAMO began to be known as the party that championed free and fair elections and adjudicated electoral disputes.[32]

Nevertheless, as relative peace in Mozambique continued and the international community turned its focus to other conflicts, RENAMO returned to its previous practices, reestablishing historic ethnic cleavages and reverting to a structure of centralized and personalized power in party President Dhlakama. In fact, instead of relying on proper administrative channels, Dhlakama is known to dismiss RENAMO officers or parliamentary delegates who gain too much power and visibility or who could pose a challenge to his authority. Dhlakama's autocratic governing style has eroded the group's initial commitment to democratic politics, made the party less effective in parliament, and reduced its chances of becoming the majority party.[33] As a result, the party has lost gains made in previous elections, losing 27 legislative seats and maintaining only 90 of 250 seats in the 2004 election.[34]

In June 2014, the group renounced its truce with the FRELIMO-dominated government, insisting that promises made by the government had not been fulfilled. Dhlakama plans to participate in the 2014 elections but publicly ended the 1992 cease-fire, stating that the FRELIMO forces reneged on the Rome Accords.[35] However, it has been suspected that the resurgent violence could be a strategy to regain some of the attention RENAMO received in the early 1990s and to avoid irrelevance in the eyes of the international community.[36]

RENAMO gained political representation as a direct result of its ability to leverage relationships with the major players in the international community to obtain financial aid. The leadership of RENAMO was willing to accept transformation into a political entity, despite facing continued bloodshed and elimination by FRELIMO forces. The negotiations that followed from the Rome Accords allowed RENAMO to become a significant player in Mozambican politics and gain de jure authority through political representation in the national legislature. The resurgence of violence came twenty years after initial funding, and observers speculate that a lack of sponsored funding has undermined the ability of RENAMO to maintain a de jure role in governance.

INTERNAL ORGANIZATION AND ADMINISTRATIVE STRUCTURE

Finally, in addition to mass support and relationships with internal and external donors, the last factor contributing to a resistance group's transition to governance is the development of an internal administrative structure to manage the provision of goods and services and oversee finances. Having a highly capable administrative capacity allows the resistance group to demonstrate its capability to the populace, organize a legitimate political arm for resistance efforts, and offer a road map and structure for a nascent government.

Eritrean People's Liberation Front

An example of a highly developed internal administrative structure within a resistance group is the Eritrean People's Liberation Front, or EPLF, which fought to form an independent state from Ethiopia. The EPLF originated from the fighting factions of another resistance group, the Eritrean Liberation Front (ELF), but eventually formed a coalition resistance movement with the ELF and other rival groups.[37]

This unification produced a political and organizational structure, headed by an administrative committee that was based on democratic centralism and obedience to an agreed-upon platform. The party also provided military training, basic education, and health benefits for its fighters. The ultimate goal of the unification was to remove any of the elements of rival factions incompatible with the movement or capable of conspiring for power.[38]

In 1978, the EPLF began organizing in towns in and near the city of Asmara to stop Ethiopian reprisals against the Eritrean liberation movements. After capturing this region militarily, the EPLF exercised de facto authority by providing health care for peasants and continuing to provide general education and military training. The EPLF also elected a central committee and established both a political bureau as well as departments for economics, health, social welfare, and mass administration, to govern Eritrean society in liberated, semi-liberated, and occupied areas.

Finally, in 1991, the EPLF took control of the city of Asmara, formed a provisional government, and attained de facto independence from Ethiopia. Ethiopia conceded to Eritrea's military victory and agreed to hold a UN-sponsored national referendum on statehood for Eritrea. An overwhelming majority of the registered voters in the state voted for Eritrea's independence, prompting the UN to recognize Eritrea's independence in 1992.[39]

After Eritrea gained independence, the EPLF transformed into the People's Front for Democracy and Justice (PFDJ) and created the Constitutional Commission for Eritrea, or CCE. This fifty-member council, composed of legal experts, academics, and PFDJ members, created a national assembly and an executive branch and established political and religious freedoms. It also decided to remove party politics by allowing only one legal political party in the country, the PFDJ, and granting the executive branch a wide latitude of powers. With only one political party in power, the central committee and elected regional delegates of the PFDJ became the legislative branch of government, which appointed the head of PFDJ, Isaias Afewerki, as president. Moreover, former fighting brigades of the EPLF became the country's military, and former war supporters were employed by ministries to facilitate the rebuilding of factories, restoration of electricity and water supplies, reconstruction of roads, and restoration of the agricultural sector.[40]

PFDJ is still the party in power, with President Afewerki maintaining an autocratic hold over the governing authority. In fact, although the constitution was ratified in 1997, the country has never held general elections for the office of the president, and the national legislature has not convened since 2002.[41]

From early in its inception, EPLF's internal hierarchy and administrative structure enabled the organization to control the constitutional process, present a united front in the war with Ethiopia, and subsequently transition its party leaders into governmental office. The PFDJ has since remained in power, drawing on the tactics that enabled it to create a unified front, including increasing centralization under the executive and undermining competition or constitutional challenges to power.[42]

Sudan People's Liberation Army

A second case in which administrative capacity facilitated the transition from de facto control to governing authority is the Sudan People's Liberation Army (SPLA). The SPLA was a resistance group of former Sudanese military fighters who sought regional autonomy for southern Sudan from the northern-led Sudanese government. Regional autonomy was awarded to the south in a peace agreement in 1983, although shortly afterward northern Sudan violated the terms of the agreement and began appropriating southern natural resources, such as water and oil. The result was a twenty-year civil war between the two regions.

This conflict caught the attention of the international community because each party committed atrocities including discrimination, village displacement, disruption of UN peacekeeping efforts, indiscriminate bombardment, creation of famine conditions, employment of child soldiers, and extreme repression.[43] However, in the late 1980s, the SPLA began cooperating with UN relief agencies to gain international credibility and protect the civilian population. This cooperation included declaring a cease-fire in designated areas that would permit famine relief agencies to conduct operations safely. By the 1990s, the SPLA gained de facto control in the southern region and claimed jurisdiction over the 4.1 million people residing in the south.[44]

After more than twenty years of devastation, the government of Sudan and the SPLA agreed to an internationally facilitated Comprehensive Peace Agreement in 2005, under which they would negotiate shared governance and oil revenues. Soon afterward, SPLA President and Commander-in-Chief Dr. John Garang established a conventional army structure to organize the more than 140,000 personnel under his command. He appointed a chief of general staff and four deputy chiefs of general staff in the areas of administration, operations, logistics, and political and moral orientation. The SPLA also divided its army into divisions and brigades to monitor jurisdictions in southern Sudan.[45]

Politically, the organization forged links with a prominent group of northern political parties sympathetic to its aims, known as the National Democratic Alliance, as well as southern rivals, the Sudan People's Defense Forces/Democratic Front and the Lord's Resistance Movement, to form the Sudan People's Liberation Army/Movement.[46] The SPLA then reorganized its structure by promoting its deputy chiefs of general staff to lieutenant generals and incorporating new officers from rival organizations into joint integrated units to avoid infighting among rival groups.[47]

During the final negotiation under the Comprehensive Peace Agreement, southern Sudan was permitted to hold a referendum for independence from the north.[48] The referendum was held in 2011, and 98.83 percent of the population voted for independence. As soon as the UN recognized South Sudan's independence, SPLA authorities began to transition into the country's administrative apparatus. For instance, John Garang's successor, Salva Kir, became president of the government of South Sudan and appointed his deputies to fill newly created cabinets such as the Ministry of Finance and the Ministry of Defense and Veterans Affairs.[49]

The decade-long transformation of SPLA's guerrilla fighting campaign into an organized army and then into the governing apparatus gained international approval for the party, allowing the administration to seamlessly transition into the offices of the government of South Sudan uninterrupted. A challenge to this administrative structure occurred in 2013, however, when the SPLA-dominated government faced attacks by groups associated with President Kir's former deputy, Riek Machar, on the nation's oil fields, although parties are now working toward a negotiated settlement.[50]

CONCLUSION

The preceding six case studies illustrate the factors that contribute to a resistance group's transition into a de facto authority, a party with national political representation, or a state governing authority. These factors are a base of support, a relationship with internal and external donors, and an administrative structure or internal organization. Although these factors aid in facilitating a move from resistance to governance, they are not necessarily sufficient for sustainable or successful governance. In the case of RENAMO, many of the reforms that facilitated the organization's transition to governance are now being undone. Furthermore, although EPLF and SPLA were successful in attaining statehood, they still struggle to maintain sustainable control, with the government of Eritrea increasingly being condemned for its violent tactics to assert governmental authority and the government of South Sudan facing domestic resistance from rebel movements. A final example, describing the contemporary resistance group in Libya, will illustrate the complications associated with transitions to governance faced by most resistance groups.

Anti-Qadhafi Resistance Forces

The anti-Qadhafi resistance forces were successful in achieving their aims of overthrowing the standing government and becoming the governing authority in Libya, but they have been unable to transform this momentum into becoming a stable ruling authority. The anti-Qadhafi forces formed in 2010, when Libyan leader Muammar Qadhafi responded to a number of civilian protests with brutal state force, killing fourteen protestors.[51] Dissatisfied with forty years of repressive rule, anti-Qadhafi protesters and demonstrators around Benghazi began overwhelming government forces and driving Qadhafi loyalists out of the city.[52]

By February 2011, it was evident that what had started as civilian protests had fomented into an uprising to remove Qadhafi from power. Qadhafi's own ministers began resigning from their posts, condemning the excessive use of force on protestors.[53]

In late March, the UN imposed a no-fly zone in Libyan airspace to stem some of the violence imposed by Qadhafi, while Great Britain, the United States, France, Qatar, and Turkey announced support for the resistance force, supplying civilians with communications equipment, training, and arms. Soon after, the international community made Libya's previously frozen assets available to the anti-Qadhafi forces, which then transitioned into a political party known as the National Transitional Council (NTC).[54]

The NTC issued a Constitutional Declaration as "the sole representative of Libya."[55] The NTC's executive board, chaired by Mahmoud Jibril, set a time line for interim governance until the liberation of Tripoli, the capital.[56] After the capture of Tripoli in August 2012, the NTC transitioned into the General National Congress (GNC), a legislative body made up of more than two hundred members who appointed Ali Zeidan as prime minister and Mohammed Magariaf as president.[57]

However, since the successful overthrow of the incumbent government, the government has struggled to rebuild much of the national infrastructure and security apparatus, culminating in the 2012 attack on the US embassy in Benghazi. Moreover, the volatile security situation in the country has produced numerous armed bands, militias, and tribal groups working to gain territorial control as well as resistance groups that aim to control Libya's vast oil reserves.[58]

Even though the NTC granted amnesty to those who committed crimes during the revolution, civilians, independently and in organized militias and tribal groups, remain well armed from fighting the regime and attack prosecutors, judicial police, and judges to prevent stable rule of law in the country. Furthermore, the freedoms, including freedom of speech and religion and women's rights, promised by the NTC and GNC have yet to come to fruition as armed civilians and groups threaten the progress of elected officials, international activists, and journalists.[59]

To date, the GNC is still working with the United States, European Union, and UN on border assistance and state building focused on training an army and police force.[60] Libya's current challenges stem in part from the fact that the anti-Qadhafi forces became united only to remove Qadhafi from power. Once this occurred, the anti-Qadhafi forces formed a government with de jure authority and international approval but without a popular base of support for a party or centrally organized administrative apparatus under clearly defined leadership. Unlike the EPLF, the GNC is now faced with consolidating a myriad of parties, interests, and allegiances, which poses a substantial challenge to the creation of a stable and sustainable government in Libya.

The failures of governance in Libya demonstrate that transition from resistance group to governing authority is a complex process. Even if the resistance group is successful in building a base of support, cultivating a relationship with donors, and gaining de jure authority to govern, the absence or weakening of any one contributing factor can undermine a resistance movement's transition to governance or its ability to govern after a conflict. The anti-Qadhafi forces initially possessed the contributing factors to some degree (mass support for the anti-Qadhafi forces and NTC, relationship to donors, and attempts at internal organization) and were successful in gaining de jure authority, but they have since been unable to maintain a consolidated base of support and transform de jure authority into sustainable governance practices.

Although there is no single road map for successful transition of a resistance group to governance, the factors identified in this paper provide insight and may increase prospects for a successful transition. Future research could focus on understanding the weight or importance of each factor or ascertaining whether the presence of certain factors over others increases the likelihood of the success and sustainability of the transitioned de jure authority.

Table 1. Summary of Resistance Groups' Transition to Governance

Resistance Group	Stage of Governing Authority	Primary Contributing Factor of Transition to Governance	Current Status
Hamas	De facto governance in the Gaza Strip/ political representation in national legislature	Domestic base of support	*Active*—Controls Gaza Strip and has representation in a national unity government (power-sharing agreement with Fatah)
Taliban	De facto governance/ de jure (in law) governing authority	Domestic base of support	*Inactive* (as de jure authority in Afghanistan)—Dismantled by US troops in Afghanistan but active as de facto government in communities in northern Pakistan
Hizbollah	De facto governance/political representation	External relationship with donors	*Active*—De facto authority in southern Lebanon and has political representation in national legislature
RENAMO	Political representation	External relationship with donors	*Active*—Has political representation in legislature, although severely weakened since the Rome Accords
EPLF	De jure (in law) governing authority	Internal organization and administrative structure	*Active*—State governing authority in Eritrea as the PFDJ

175

Resistance Group	Stage of Governing Authority	Primary Contributing Factor of Transition to Governance	Current Status
SPLA	De jure (in law) governing authority	Internal organization and administrative structure	*Inactive*—State governing authority in the government of South Sudan and inactive as the SPLA
Anti-Qadhafi forces/NTC	De jure (in law) governing authority	Domestic base of support	*Active*—State governing authority of Libya although faces considerable obstacles and challenges to power

NOTES

1 Jonte Van Essen, "De Facto Regimes in International Law," *Utrecht Journal of International and European Law* 28, no. 74 (2012): 31–49; and Zachariah Cherian Mampilly, *Rebel Rulers: Insurgent Governance and Civilian Life during War* (Ithaca, NY: Cornell University Press, 2011), 39.

2 "Who are Hamas?" *BBC News*, updated January 4, 2009, http://news.bbc.co.uk/2/hi/1654510.stm.

3 Matthew Levitt and Dennis Ross, *Hamas: Politics, Charity, and Terrorism in the Service of Jihad* (Washington, DC: The Washington Institute for Near East Policy, 2006), 26–28; and Jonathan Masters and Zachary Laub, "Hezbollah (a.k.a. Hizbollah, Hizbu'llah)," Council on Foreign Relations, January 3, 2014, http://www.cfr.org/lebanon/hezbollah-k-hizbollah-hizbullah/p9155.

4 Scott Wilson, "In Politics, Hamas Gains in the West Bank," *Washington Post*, June 29, 2005, http://www.washingtonpost.com/wp-dyn/content/article/2005/06/28/AR2005062801368.html.

5 Levitt and Ross, *Hamas: Politics, Charity, and Terrorism in the Service of Jihad*, 51.

6 "Middle East: Why Did Hamas Win Palestinian Poll?," Radio Free Europe/Radio Liberty, January 26, 2006, http://www.rferl.org/content/article/1065113.html.

7 Wilson, "In Politics, Hamas Gains in the West Bank."

8 "Who are Hamas?" *BBC News*; and Jim Zanotti, "Hamas: Background and Issues for Congress," CRS Report no. R41514 (Washington, DC: Congressional Research Service, December 2, 2010), http://www.fas.org/sgp/crs/mideast/R41514.pdf, 3–4.

9 Ahmed Rashid, *Taliban: Militant Islam, Oil, and the New Great Game in Central Asia* (London: I. B. Tauris, 2000), 21.

10 Ibid.

11 Kenneth Katzman, *Afghanistan: Post-Taliban Governance, Security, and U.S. Policy*, CRS Report no. RL30588 (Washington, DC: Congressional Research Service, July 11, 2014), https://www.fas.org/sgp/crs/row/RL30588.pdf, 3; and Zachary Laub, "Backgrounder: The Taliban in Afghanistan," Council on Foreign Relations, updated July 4, 2014, http://www.cfr.org/afghanistan/taliban-afghanistan/p10551.

12 Rashid, *Taliban: Militant Islam, Oil, and the New Great Game in Central Asia*, 25.

13 Ibid., 29–30.

14 Katzman, "Afghanistan: Post-Taliban Governance, Security, and U.S. Policy," 5.

15 Laub, "The Taliban in Afghanistan."

16 "How the Taliban Gripped Karachi," *BBC News*, March 20, 2013, http://www.bbc.com/news/world-asia-21343397.

17 Mampilly, *Rebel Rulers*, 234.

18 Masters and Laub, "Hezbollah (a.k.a. Hizbollah, Hizbu'llah)."

19 US Department of Homeland Security, Immigration and Customs Enforcement, "Mohamad Youssef Hammoud Sentenced to 30 years in Terrorism Financing Case," news release, January 27, 2011, http://www.ice.gov/news/releases/1101/110127charlotte.htm.

20 Shana Marshall, "Hizbollah: 1982-2009," in *Casebook on Insurgent and Revolutionary Warfare, Volume II: 1962–2009*, ed. Chuck Crossett (Fort Bragg, NC: United States Army Special Operations Command, 2012), 552.

21 Judith Palmer Harik, *Hezbollah: The Changing Face of Terrorism* (London: I. B. Tauris, 2004), 60–62.

22 "Lebanon: The Many Hands and Faces of Hezbollah," *IRIN*, March 29, 2006, http://www.irinnews.org/report/26242/lebanon-the-many-hands-and-faces-of-hezbollah.

23 Masters and Laub, "Hezbollah (a.k.a. Hizbollah, Hizbu'llah)"; and Marshall, "Hizbollah: 1982–2009," 560.

24 Iain Scobbie, "South Lebanon 2006," in *International Law and the Classification of Conflicts*, ed. Elizabeth Wilmshurst (Oxford: Oxford University Press, 2012), 403.

25 James M. Scott, *Deciding to Intervene: The Reagan Doctrine and American Foreign Policy* (Durham, NC: Duke University Press, 1996), 194–195.

26 Ibid., 197.

27 "Terrorist Organization Profile: Mozambique National Resistance Movement," National Consortium for the Study of Terrorism and Responses to Terrorism, http://www.start.umd.edu/tops/terrorist_organization_profile.asp?id=314.

28 Christoph Zürcher, Carrie Manning, Kristie D. Evenson, Rachel Hayman, Sarah Riese, and Nora Roehner, *Costly Democracy: Peacebuilding and Democratization after War* (Stanford: Stanford University Press, 2013), 104.

29 Carrie Manning, "Armed Opposition Groups into Political Parties: Comparing Bosnia, Kosovo, and Mozambique," *Studies in Comparative International Development* 39, no. 1 (2004): 54–76.

30 Carrie Manning, *The Making of Democrats: Elections and Party Development in Postwar Bosnia, El Salvador, and Mozambique* (New York: Palgrave Macmillan, 2008), 48–53.

31 "Terrorist Organization Profile: Mozambique National Resistance Movement," National Consortium for the Study of Terrorism and Responses to Terrorism; Gwinyayi Dzinesa and Paulo Wache, "Mozambican Elections: What to Make of Dhlakama's Intention to Run for President," *All Africa*, May 27, 2014, http://allafrica.com/stories/201405270478.html; and Manning, "Armed Opposition Groups into Political Parties."

32 Manning, *The Making of Democrats*, 56.

33 Manning, "Armed Opposition Groups into Political Parties."

34 Manning, *The Making of Democrats*, 51.

35 Foreign Staff, "RENAMO suspends Mozambique Ceasefire, Vows to Step Up Attacks," *Business Day Live*, June 6, 2014, http://www.bdlive.co.za/africa/africannews/2014/06/06/renamo-suspends-mozambique-ceasefire-vows-to-step-up-attacks.

36 Reuters, "Mozambique Army Launches Retaliatory Attack on RENAMO," *Yahoo Finance UK & Ireland*, October 18, 2013, https://uk.finance.yahoo.com/news/mozambique-army-launches-retaliatory-attack-183944138.html.

37 David Pool, *From Guerrillas to Government: Eritrean People's Liberation Front* (Athens, OH: Ohio University Press, 2001), 55.

38 Ibid., 75–76.

39 Ibid., 161.

40 Ibid.,163.

[41] Tesfa-Alem Tekle, "Eritrean Leader Pledges to Draft New Constitution," *Sudan Tribune*, May 24, 2014, http://www.sudantribune.com/spip.php?article51115.

[42] "World Report 2013: Eritrea," Human Rights Watch, January 1, 2013, http://www.hrw.org/world-report/2013/country-chapters/eritrea.

[43] Jemera Rone, John Prendergast, and Karen Sorensen, *Civilian Devastation: Abuses by All Parties in the War in Southern Sudan* (New York: Human Rights Watch, 1994), 2–3.

[44] Charles Tripp, "The Sudanese War in International Relations," in *After the Cold War: Security and Democracy in Africa and Asia*, eds. William Hale and Eberhard Kienle (London: Tauris Academic Studies, 1997), 57.

[45] Richard Rands, "In Need of Review: SPLA Transformation in 2006-2010 and Beyond," *Small Arms Survey*, HSBA Working Paper 23, November 1, 2010, 19–20, http://www.smallarmssurveysudan.org/fileadmin/docs/working-papers/HSBA-WP-23-SPLA-Transformation-2006-10-and-Beyond.pdf.

[46] Claire Metelits, *Inside Insurgency: Violence, Civilians, and Revolutionary Group Behavior* (New York: New York University Press, 2010), 68.

[47] Rands, "In Need of Review: SPLA Transformation in 2006-2010 and Beyond," 21.

[48] "Sudan: Southern Pull-out Threatens Peace Deal," *IRIN*, October 11, 2007, http://www.irinnews.org/report/74746/sudan-southern-pull-out-threatens-peace-deal.

[49] Patrick Worsnip, "South Sudan Admitted to U.N. as 193rd Member," *Reuters*, July 14, 2011, http://uk.reuters.com/article/2011/07/14/uk-sudan-un-membership-idUKTRE76D3I120110714.

[50] Nicholas Bariyo, "South Sudan Peace Talks Reach Apparent Breakthrough," *Wall Street Journal*, June 11, 2014, http://online.wsj.com/articles/south-sudan-peace-talks-reach-apparent-breakthrough-1402473616.

[51] "Protesters Take Control of Several Libyan Cities," *The News International*, February 19, 2011, http://www.thenews.com.pk/TodaysPrintDetail.aspx?ID=31907&Cat=1.

[52] Anne Barker, "Time Running Out for Cornered Gaddafi," *ABC News*, February 24, 2011, http://www.abc.net.au/news/2011-02-24/time-running-out-for-cornered-gaddafi/1955842.

[53] Al Jazeera and Agencies, "Gaddafi Loses More Libyan Cities," *Al Jazeera English*, February 24, 2011, http://www.aljazeera.com/news/africa/2011/02/2011223125256699145.html; and "Libya: UN backs Action Against Colonel Gaddafi," *BBC News*, March 18, 2011, http://www.bbc.co.uk/news/world-africa-12781009.

[54] "Libya Conflict: UN Agreement Sought to Unfreeze Assets," *BBC News*, August 25, 2011, http://www.bbc.co.uk/news/uk-politics-14661504.

[55] "Founding Statement of the Interim Transnational National Council," Lauterpacht Centre for International Law, March 5, 2011, http://www.lcil.cam.ac.uk/sites/default/files/LCIL/documents/arabspring/libya/Libya_12_Founding_Statement_TNC.pdf.

[56] Vivienne Walt, "How Did Gaddafi Die? A Year Later, Unanswered Questions and Bad Blood," *Time*, October 18, 2012, http://world.time.com/2012/10/18/how-did-gaddafi-die-a-year-later-unanswered-questions-and-bad-blood/.

[57] Rana Jawad, "Libyan Voters Prepare for Change," *BBC News*, July 5, 2012, http://www.bbc.com/news/world-africa-18721576; and Ali Shuaib, "Libyan Assembly Votes Gaddafi Opponent as President," *Reuters*, August 9, 2012, http://www.reuters.com/article/2012/08/09/us-libya-assembly-idUSBRE87811D20120809.

[58] "World Report 2014: Libya," Human Rights Watch, http://www.hrw.org/world-report/2014/country-chapters/libya, 1.

[59] Ibid., 3.

[60] Ibid.

INTERNAL CONFLICT AND THE ROLE OF OUTSIDE INFLUENCE: HOW FOREIGN STATES LEVERAGE INDIGENOUS RESISTANCE MOVEMENTS TO ACHIEVE NATIONAL OBJECTIVES

Joseph M. Tonon, Erin N. Hahn, and Summer Newton

INTRODUCTION

Since the Cold War, external support of resistance movements has been a common method for states to achieve various strategic political objectives, such as destabilizing rivals, increasing regional influence, or instituting regime change. Working through indigenous populations, states have extended their ability to influence the outcomes of conflicts without having to engage directly with rival states. However, since the end of the Cold War, the nature of internal conflict has dramatically changed and, accordingly, so too has the type and scope of outside support. In particular, the amount of funding and resources outside actors provide to resistance movements has lessened, yet the impact and effectiveness of this support has not. The purpose of this paper is to provide an introductory understanding of the conditions that give rise to internal conflict as well as the strategies that outside actors, particularly sovereign states, use to shape the conflict or pursue other objectives through support of a resistance movement. In particular, this paper looks at the following questions:

1. Why and in what ways do outside actors provide external support to resistance movements to influence an internal conflict or achieve other objectives?

2. How has the nature of conflict changed since the end of the Cold War, and what implications does that have on external support?

The context in which one country may attempt to influence outcomes in another by employing indigenous populations is important to understand, particularly, from the United States' perspective with regard to unconventional warfare (UW). The US military defines UW as "activities that are conducted to enable a resistance movement or insurgency to coerce, disrupt, or overthrow a government or occupying power by operating through or with an underground, auxiliary, and guerrilla force in a denied area."[1] The ideas and concepts outlined in the current definition of UW are neither simple nor easy to implement. There are many implications that follow from the definition that have yet to be explored. However, to speak of external support from a US view is to speak of UW. This point matters because Special Operations Forces (SOF) personnel charged with carrying out UW need to understand how resistance movements operate, which requires the ability to recognize the agendas and forms of support provided by external states, as well as the relevance of other external actors (e.g., diasporas and refugee populations). Moreover, the dynamics of external support are important to understand not only so that the United States can be more effective at carrying out UW but also to ensure that SOF personnel understand the effects of such support on groups the United States seeks to counter.

This paper was developed as an initial exploration of the topic of external support, with the goal of highlighting both relevant theoretical approaches and cases for more in-depth follow-on study. It provides an overview of key elements, including the forms of and motivation for post-Cold-War external support. Amid budgetary constraints and an ongoing threat to national security posed by global terrorism, it is likely that the United States will need to achieve national security and strategic political objectives through means other than direct combat operations. Given its historical use and current relevance in internal conflicts around the world, UW may be a viable option for furthering these objectives. To better understand why UW may be a viable strategic option for the US government to pursue, research needs to be conducted on the reasons behind a state's decision to use an indigenous population (i.e., what theory or theories help explain state support of resistance movements, and is there theory that underpins existing UW doctrine)? Whether UW is pursued as a strategic policy option is also dependent on how it is presented and understood. What tool or tools can help explain why it is a viable option that will resonate with policy makers? Making the case for UW requires imparting a deeper understanding to nonpractitioners of what it involves and what risks are associated with its execution.

Because this paper is an introductory discussion of the topic, the overall goal is to conduct case studies over time that will permit a more in-depth understanding of the interplay of variables in external support and when it has worked or failed and why. Detailed study will lead to a better understanding of the questions posed above and to the eventual distillation of best practices, which can be used to maximize SOF personnel's understanding of UW.

It is important to note that most if not all other foreign countries do not use the term *unconventional warfare* and instead embrace the language of insurgency and terrorism. In this context, insurgency usually means waging a guerrilla war against a sitting government or occupying power. Since the end of World War II, these tactics have been employed in most inter- and intrastate conflicts. After the Cuban missile crisis, the superpowers realized that direct confrontation was too dangerous. Instead, they competed with each other on the periphery, and they used other nations in the context of internal conflicts between indigenous populations or sitting governments to do their bidding. For example, during the Angolan War of Independence, the United States indirectly supported the sitting government, while the Soviet Union supported the Popular Movement for the Liberation of

Angola, an insurgent group that was challenging the government for control of the country, with arms and other supplies. States have come to accept that it is too costly to engage in direct conflict, particularly if the political objective is important but not crucial to national security. External support is a viable option, but it comes with costs and benefits. Understanding the risks attendant to providing support is important.

The more current example in Syria underscores this point. After confirmation of the Syrian government's use of chemical weapons June 2013, the United States provided direct military support to opposition forces aimed at toppling Assad's regime. For at least two years prior, discussions encouraging US support of opposition forces swirled in Congress, and some news outlets even questioned whether Syria was the next proxy war. When President Obama did publicly authorize support, the Syrian opposition was in a somewhat fractured state. It consists of a coalition of different groups, often with very different ideologies and ideas about Syria's future. The opposition forces subsequently formed an umbrella organization, but concerns remained that the loose confederation could quickly splinter over how to handle the transition if Assad were overthrown. Of more immediate significance was the strife within the opposition forces that forced both the United States and the United Kingdom to suspend outside aid. Meanwhile, Iran, Lebanese Hizbollah supported by Iran, and Iraqi Shiite factions were very effective at training progovernment paramilitary groups and merging disparate forces to create a strong coalition keen on keeping Assad in power. What factors made Iran's support of pro-Assad groups so effective, and what factors contributed to the inability of the opposition forces to solidify? What were the US motivations for supporting the opposition forces, and to what extent were the risks borne out by later events understood in advance of the support? There are numerous factors that need to be explored to begin to answer these questions, in the Syria example and any other conflict involving significant external support. The factors are identified in this paper, and the subsequent UW case studies are intended to provide case-specific analysis of those factors.

This paper uses the term *resistance movement* to refer to any group opposing a standing government. This term was selected not only because it is used in the UW definition but also for its inclusiveness; it accounts for groups that external actors may wish to support, which may share many but not all the same characteristics. For instance, a group may be in the nascent stages of development and attract external support for that very reason. Others may be more sophisticated, participating in an insurgency, and may also be the target of external support based on their level of development and ability to help a state achieve certain ends within another sovereign territory. The terms *insurgents* and *rebels* are commonly used in the media and literature in generic fashion. Care has been taken to avoid the imprecise use of these terms here, and any reference to either should be understood as simply a specific type of resistance movement.

RESEARCH QUESTIONS

The literature we draw on for this paper is, for the most part, from economics and political science. With this literature in mind, there are two lenses through which to view this large body of work. The first is from the perspective of a group of indigenous actors trying to change or defeat a sitting government. Alternatively, the other perspective is from the outside actor, in this case some state that seeks to influence the outcome of a conflict.[a]

To begin to understand the effectiveness of external support and the best types of support for a state to provide, it is necessary to conduct a series of comparative case studies on support to indigenous resistance movements. By using the comparative method, it is possible to search for lessons that generalize but that also have enough historical context to allow the reader to adapt them to the circumstances of the particular case. By carrying out this research, it will be possible to establish a set of best practices for SOF personnel. The UW case studies along with work on correlating the phases of organizational growth with the stages of an insurgency are central to this effort and key for the SOF community as it continues to build its capacity to execute this difficult and demanding mission.

In the empirical sciences, there is always a dialogue between the development of theory to explain some type of social phenomenon and data collection and analysis. Researchers seek to strike an appropriate balance between the theoretical and the empirical. As we move forward with more-detailed case studies and use the comparative method to draw inferences, we will learn more about the types of support that are most effective and the best practices associated with their delivery.

Some of the questions that inform this work include the following:

- Why do states that are external to a conflict use indigenous forces to achieve their strategic objectives?

- Once states make the decision to use indigenous forces, what types of support are most effective?

---

[a] Three very important books in this area are Kalyvas's *The Logic of Violence in Civil War*, Petersen's *Resistance and Rebellion*, and Weinstein's *Inside Rebellion*.

- If you help create a capacity for societal change, violent or otherwise, can it be controlled?
- What types of mechanisms help with control?
- If a group receives assistance from outside a state's borders, does this change the government's response?
- Does outside support during an internal conflict lengthen the duration of the conflict?

These questions helped to frame our guidelines and criteria, which are discussed in the "Potential Case Studies and Case Selection" section.

CIVIL WAR DYNAMICS IN THE POST-COLD-WAR ERA

External-state supporters are providing resources, whether material, expertise, or other forms of support, to resistance movements opposing a standing government or occupying power. Gaining a clearer understanding of how such support can be an effective means of achieving national objectives requires a close look at the overall context in which it takes place. Not all instances of external support, or all historical periods in which it has taken place, are equal. Examining external support for resistance movements during the Cold War is an important component of understanding the practice. Yet, the demise of communism in the Soviet Union and Eastern Europe marked a watershed event that altered the conditions in which external support, and armed conflict, takes place. In many ways, the basic motivations that induce people to take up arms are enduring. Many do so to protect themselves and their families, to gain control of scarce resources, or to have more control over the future of their lives and community. However, patterns of armed conflict have changed considerably in the aftermath of the Cold War. Our examination of external support must take into account the ways in which these trends alter the practice and its expected outcomes. This section discusses these emerging trends in armed conflict: the frequency of intrastate war, the frequency and type of conflict resolution, altered organizational motivations for conflict, and, lastly, changes in methods of warfare.

Academics and policy makers are able to study patterns in armed conflict from a different perspective than the soldier concerned with the conflict. These patterns, and the structures that influence them, are often difficult to see for those experiencing the conflict firsthand. This is true of other domains of complex human interaction as well, such as economics. An American unemployed after the 2008 recession, for instance, is likely to explain his financial hardship in terms close to home. Nevertheless, global commodity markets, international financial organizations, and political instability in remote regions are all factors far afield from our everyday lives that impact us intimately. Likewise, the actions of national governments, resistance movements, and even individuals are shaped by larger structures, such as the relationships between powerful states that form the architecture of the international system.

Frequency of Intrastate War

The end of the Cold War resulted in changes within the international system that impacted the dynamics of civil war the world over. During the Cold War, the rivalry between two superpowers, the United States and the Soviet Union, shaped the *when, where,* and *how* of internal conflict. The high costs of war between the two superpowers meant that the rivalry was carried out in other venues, usually in other states between combatants supported by either side.

With the end of that global rivalry, there was a noticeable uptick in the frequency of civil war, leading some to fear a period of coming anarchy. The sharp spike in the number of civil wars peaked in 1992, but many of those conflicts were contained by 1996.[2] The frequency of interstate wars continued to drop. Today, most armed conflicts take place within states. In 2011, for instance, of the thirty-seven active conflicts (defined as those resulting in more than twenty-five battle deaths per annum) worldwide, only one of those conflicts was between states.[3] During the Cold War, many civil wars were long-lived and difficult to resolve, most likely because of superpower support. Beginning in the 1990s, by contrast, civil wars, many of which had begun in previous decades, resolved at a rapid rate. In the forty-five-year period of the Cold War, for example, only 141 intrastate conflicts ended. By contrast, in the short fifteen-year period after the end of the Cold War, 145 intrastate conflicts ended.[4]

Conflict Resolution

Not only are more civil wars finding resolution today, but the manner in which civil wars end has also experienced a sea change. Beginning in 1940 up until the end of the 1980s, most civil wars ended via a decisive military victory, whether by rebel forces or the incumbent government. From 1940 to 1999, civil wars were four times as likely to end via a military victory as a negotiated settlement. However, beginning in 1990, not only did an astonishing number of civil wars (many of which had begun in the Cold War era) end, but many of them ended not through a military victory but by a negotiated settlement between the opposing sides. In the 1990s, only four out of every ten civil wars ended through a military victory. Of the thirty-seven civil wars that ended in the 1990s, fifteen ended in a military victory, fifteen by negotiated settlement, and seven by a cease-fire/stalemate.[5] When the dataset used to calculate those figures is extended to include low-intensity conflicts, the difference is even more striking. In the period 1990–2005, of the 147 conflicts recorded, twenty-seven terminated through a negotiated settlement, whereas only twenty terminated via a military victory.[6,b]

Statistics reveal that not all resolutions to conflict are equal. The higher rate of negotiated settlements has left more room for the international community to influence outcomes as third-party states or international organizations help mediate settlements and keep the peace afterward. Despite the intervention by the international community, civil wars ended through such settlements are twice as likely to reignite as those resolved via a decisive military victory. Surprisingly, the most stable outcome of a civil war has proven to be a clear military victory by rebel forces.[7]

What about negotiated settlements makes them so precarious? These settlements are good at distributing important resources, such as political offices, making it easier to bring opposing combatants to the bargaining table. However, there is often no third party able to make certain that each side upholds its end of the bargain. Countries face many challenges in the aftermath of civil war. One political scientist, Monica Toft, has researched the instability of negotiated settlements extensively.[8] She finds that the most important task for ensuring stability after a settlement is good security-sector reform. Combatants from both sides, whether government or rebel forces, must be well integrated and managed. However, the international community often perceives this task as being less pressing than those of humanitarian relief or the prosecution of war crimes.

---

[b] Kreutz uses the UCDP-PRIO dataset that includes conflicts resulting in more than twenty-five battle-related deaths per annum. Monica Duffy Toft, by contrast, uses the Correlates of War dataset, which only tracks conflicts that result in one thousand or more battle-related deaths per annum. As an example of how Toft's higher threshold affects the dataset, two successful insurgencies, the April 19th Movement (based in Colombia), or M-19, and the Provisional Irish Republican Army (based in Northern Ireland), or PIRA, are both excluded from the Correlates of War dataset. Both groups, incidentally, terminated their armed struggle after reaching a negotiated settlement with the incumbent government.

Conflict Motivation

The end of the international system structured by the Cold War has also altered the motivations for conflict. Certainly, the incidence of insurgencies fueled by communist-inspired ideologies decreased, especially in Latin America. Conflicts fought over issues of ethnic identity, by contrast, increased. Another theory describes armed conflict motivated more by "greed" than "grievance."[9] According to this understanding of armed conflict today, insurgents are no longer ideologically inspired rebels struggling for justice but rather opportunistic groups whose primary goal is personal enrichment. Now, most agree that this trend is most often found in Sub-Saharan Africa.[10] Another theory contrasts current "new wars" with "old wars" fought during the Cold War. The new wars prevalent today, proponents of this theory argue, are more likely to be fought by irregular forces and motivated more by issues of identity than geopolitical interests or ideologies. These forces are also more likely to target civilians than enemy combatants. Lastly, resources fueling the fight derive from more unconventional sources, such as from loot or taxes on humanitarian aid.[11]

Methods of Warfare

One last important shift in the post-Cold-War era relates to changes in the methods by which armed conflicts are typically fought today. After the dissolution of the Soviet Union, conflicts erupted in its former republics and buffer states in Eastern Europe. The antagonists in these wars were usually competing factions of former government forces. The conflict dynamics in these countries, as well as others, have increased the number of conventionally fought wars after the Cold War. Irregular or guerrilla warfare[c] was the most common method of rebellion during the Cold War. It emerged as "people's war" in 1930s with the help of Mao. The techniques were further honed by communist resistance movements in Europe and Asia during World War II. Soon after, the methods migrated to the developing world. Irregular warfare proved remarkably effective. Previously, conventional armies were most

---

[c] Stathis Kalyvas distinguishes irregular or guerrilla warfare as asymmetrical conflict between states and rebels. The methods adopted by the guerrillas include "indirect and low-level engagement" with a high frequency of ambush and raid.[12]

often able to defeat irregular armies.[d] During the Cold War, resistance movements using irregular warfare methods were more likely to defeat conventional forces than ever. A mixture of factors contributed to this success, including external support from superpowers. The Soviet Union was especially adept in this regard. Its provision of materials, revolutionary ideology, and military doctrine helped many resistance movements mount successful campaigns against standing governments.[14]

One of the consequences of the end of the Cold War is weaker states. During the Cold War period, funds regularly flowed from the coffers of the United States and the Soviet Union to prop up weak regimes. When those states stopped receiving support, many no longer had the same capacity to exert sovereign authority over their territory. These weak states are especially prevalent in Sub-Saharan Africa. Not surprisingly, many of the conflicts fought after the Cold War have been in regions with a surfeit of weak states. In these conflicts, weak challengers fight against weak states. The rebel and government forces fighting in these weak states are typically equally militarily matched—albeit poorly.

The quintessential model of irregular warfare is a weak challenger against a strong state, but this pairing does not occur as frequently as it used to. What is not yet understood is whether these conflicts, which pair weak states with weak challengers, are the same animal as the typical irregular conflict pairing strong states with weak challengers. One academic thinks this is a new method of warfare, substantively different from irregular warfare. If he is right, then according to his research, the incidence of what he calls "symmetrical nonconventional," or SNC, warfare is the most frequent method of warfare today.[15] This is contrary to the prevailing wisdom of the Department of Defense, which would cite irregular conflict as the most frequent method of warfare. According to this research, the incidence of irregular conflict has decreased from its heyday in the Cold War. Thus far, overall no smoking gun points to a significant difference between so-called SNC and irregular methods

---

[d] One fascinating study describes why conventional armies were so successful in combating irregular armies in the nineteenth century as compared with the twentieth century. In the nineteenth century, militaries were "foraging" armies, requiring a great deal of interaction with local populations while gathering necessary supplies in conflict zones. As a result, they were able to acquire more information from local populations, helping them to more easily tell the difference between combatants and noncombatants. The increasing mechanization of Western armies in the twentieth century made them less dependent on local populations for survival, which, in turn, made them less effective at counterinsurgency. In addition to examining a large dataset of historical insurgencies, the researchers also use case studies on the intelligence-gathering capabilities of two US Army divisions in Iraq to support their conclusions. Most research on the impact of technology focuses on its capacity to create a level playing field in asymmetric conflict. This research highlights some unusual, and much less discussed, second- and third-order effects of technology on warfare.[13]

of warfare, but due attention should be given in future research on conflicts fought between weak challengers and weak states.

Implications for External Support

The impact of these trends on the practice, and utility, of external support to resistance movements is not fully understood. Transitioning a resistance movement to governance in the wake of a military victory is a difficult endeavor regardless of the circumstances. The peace is precarious as members of the former resistance movement struggle to transition from soldiers to politicians, build legitimacy among the population, and contain or integrate former government combatants. Current research tells us that these processes only become more challenging when conflict resolution occurs as part of a negotiated settlement. Close attention to security-sector reform is especially vital. Likewise, most external support no longer occurs in the premade ideological framework that pitted liberal democracy against socialism and communism. Only radical Islam has managed to construct comprehensive political and social visions that are compelling on a global scale. The component of resistance movements in the Arab Spring touting liberal democratic ideals—and their external supporters such as those in Egypt, Libya, and Syria—failed to meaningfully influence the political trajectory of those countries.

As of yet, we do not have a good understanding of how the method of warfare and the phase of the resistance movements affect the scope, strategies and tactics, and tenor of external support. External support provided in the incipient phase will of necessity differ from support provided to a full-blown militarized resistance movement like Hizbollah. Likewise, not all resistance movements are engaged in combat with strong government forces. Some of these conflicts are conventional, whereas others, mostly fought in weak states, are SNC. Much of what we know about external support applies primarily to support for irregular forces.

Lastly, in the absence of robust superpower support, resistance movements and their external supporters rely on unconventional revenue streams, sometimes by methods adverse to local populations, such as looting or organized crime. Many movements also take advantage of the significant amounts of humanitarian aid flowing into conflict zones. Academics and policy makers are only now beginning to shed light on how well-intentioned humanitarian aid impacts the duration and severity of conflicts. Perhaps the trend that overshadows even these concerns is the track record of nonviolent movements. Throughout the last century, it has not been force of arms that most often brought down despots or reformed corrupt government practices, but peaceful organized resistance.[16]

TYPES OF ACTORS PROVIDING
EXTERNAL SUPPORT

External support can dramatically impact the development and effectiveness of a resistance movement. Both state and non-state support can be critical to a group and may serve to increase the intensity and duration of the conflict[17] or even cause a civil conflict to become internationalized.[e] The nature of the support is related to the external actor providing the support. During the Cold War, state actors provided the majority of external support to resistance movements, often in the form of generous financial backing or the broad provision of other resources.[19] However, the characteristics of conflict have changed since the end of the Cold War, and so have the actors providing external support. This paper discusses four main categories of actors that provide support to resistance movements: states, diasporas, refugees, and other non-state supporters such as religious organizations, nongovernmental organizations (NGOs), and affluent individuals, among others.[f]

The categories of non-state supporters (e.g., diasporas and refugees) are important because of their relevance starting at end of the Cold War, when "state support [was] no longer the only, or necessarily the most important, game in town."[21] Although non-state support may prove crucial to a resistance movement, particularly in the early stages of its development, state support is the focus of this and subsequent related studies because of the nature and scope of aid that external states can provide throughout a movement's life span. Nonetheless, it is important to understand other influential categories of actors and in what ways they may influence a resistance movement through external support.

Who provides support and why can vary significantly across resistance movements. The four categories of actors each offer specific

---

[e] For an explanation of how external support can internationalize a conflict, see Hahn and Lauber.[18] One example is the 2011 conflict in Libya, where a United Nations Security Council resolution established a no-fly zone and permitted outside states to offer support to the opposition in order to protect civilians. Even though foreign troops were not allowed to occupy Libyan territory pursuant to the resolution, the introduction of such significant outside support arguably created an international armed conflict.

[f] Byman et al.[20] identify the categories that are used here. Within the category of other non-state support, their report looks at religious organizations, other revolutionary or insurgent groups, human rights organizations, wealthy individuals, inspirational religious leaders, NGOs, and volunteers from neighboring states. This document will not look at these other non-state groups in detail, as their support may strengthen a resistance movement early on, but their long-term impact is not comparable to that of states, diasporas, or refugees.

types of support based on a variety of different motivations. The decision to offer external support and the decision of a resistance movement to accept it involve a cost–benefit analysis on each side.[22,g] The value of the support offered depends on a number of factors, such as the requirements of the resistance at the particular time the support is introduced and the incentives in place to accept it. The external supporter does not hold the only bargaining chip, even though it ostensibly can offer resources that appeal to the resistance movement or are even required for it to progress. Indeed, in some instances, an actor may be coerced into providing support, especially if it is part of a disenfranchised group, such as refugees.[24] The resistance movement may have reasons for attracting one external supporter over another, particularly if acceptance would increase the amount of influence the supporter has over the group's major operational decisions. The actors' motivations to provide support, and the forms of support they offer (e.g., money, arms, personnel), are discussed in more detail in the "Motivations for External Support" and "Forms of External Support" sections, respectively. This section provides a brief overview of the characteristics of the actors, recognizing that the motivations and forms of support are intertwined with the identity of the actor.

State Support

A state actor refers to a recognized sovereign power. As previously noted, the Cold War rivalry between the United States and the Soviet Union is a classic example of state support, where the two superpowers and regional partners freely provided money to their preferred resistance movements. At that time, states were the primary actors providing outside aid. Although the amount of aid states furnish to groups has diminished in the post-Cold-War era, their level of support has still had a profound impact on many resistance movements in subsequent years.[25] Examples of key state support include the Ugandan and Rwandan support of the Congolese Rally for Democracy (CRD) and a corollary group, the Movement for the Liberation of the Congo (MLC) in

---

g Salehyan, Gleditsch, and Cunningham discuss this dynamic in the context of the theory of supply and demand.[23] The supply side has determinants that may explain why an external supporter identifies a particular group, including the motivations and trade-offs. For example, an external state may delegate conflict to a rebel group if the state seeks to avoid international condemnation and if the interests at stake are not critical to the state. The state can afford to lose some control inherent in delegation if its interests are not critical to state security or cannot be better served in another way. The demand-side determinants are those factors that may explain why a resistance movement seeks or accepts external support. For example, a rebel group may need funding or arms but must weigh acceptance of such support against the loss of autonomy that occurs when beholden to an outside patron.

the late 1990s.[h] The CRD and MLC were not popular locally and lacked resources, making the outside state support a critical factor in their formation and existence, even though the scale of the support did not rival the levels common during the Cold War.

Diaspora Support

A diaspora refers to an immigrant community established in a foreign country. Diasporas often continue to identify themselves with the country from which they came, which in some cases can manifest as a powerful sense of obligation to their homeland. This tie to a homeland may serve as the primary motivation for providing external support to a group. Diaspora support has played a significant role in sustaining insurgent and other groups, and the support is often related to resistance movements catalyzed by ethnic conflict. In fact, diaspora support in the form of money, arms, and recruits has occurred in every region of the world except Latin America.[27] The relevance of diaspora support to resistance movements has increased, particularly in ethnic or ideological conflicts, in part due to the diminished, albeit still important, role of state support. For example, the population of ethnic Tamils in India supported the Liberation Tigers of Tamil Eelam (LTTE) against the Sinhalese regime in Sri Lanka. The LTTE is recognized as an insurgency that was extremely effective at integrating significant diaspora support into a larger global support system.[28]

Refugee Support

As used here, *refugees* refers to displaced populations living outside of their home country or territory. Refugees have been important supporters of resistance movements largely because the violence or discrimination that caused their displacement fuels their desire to contribute to the conflict and regain the ability to return home. Unlike states or diasporas, refugees are often living in substandard conditions and are therefore unable to provide support in the form of funds or weapons. Instead, they more commonly provide personnel by sending recruits to back a movement or by developing a resistance themselves. One example is the group of Afghan refugees displaced in Pakistan who eventually formed the Taliban. In some instances, the vulnerability of refugees may be exploited and they may be coerced into providing support to a group, as has happened when refugee camps fall

---

[h] The CRD and MLC are also discussed with additional context in the "Supported Resistance Movements" section. For a detailed discussion of the formation of these groups and the eventual split of the CRD into competing factions, see Byman et al.[26]

under the control of rebels. Refugees may require the support of the host nation where they are encamped in order to effectively back movements in their home country.

Other Non-State Support

The remaining actors who fall within this category have provided important support to resistance movements but overall have contributed less significant levels of support than the previous three categories of actor. They include religious leaders and organizations, NGOs, and, in particular, human rights groups, wealthy individuals, and a rebel or insurgent group supporting another rebel or insurgent group. In addition to providing money and volunteers, religious groups can be critical in providing nonmaterial support in the form of justification for the cause through religious rulings or interpretations of religious doctrine. Wealthy individuals may support a movement out of ethnic affinity or common ideology, or because the movement's objectives resonate or align with personal values.[i]

Unlike the other actors, NGOs may inadvertently rather than purposely support a resistance movement in order to carry out their mission. For example, relief groups may provide resources to assist individuals in need and unintentionally pass resources to a rebel or insurgent group. In some cases, NGOs may knowingly provide resources directly to the resistance movements in order to ensure that emergency supplies reach individuals in areas that are rebel strongholds. Similarly, human rights groups that call attention to a conflict's atrocities (e.g., a government's use of torture or child soldiers) may unintentionally generate support for antigovernment resistance movements that results in international assistance from groups and individuals alike.

One rebel group that is supporting another rebel group is arguably one category of actor that has been more influential than others in this category, as exemplified by Charles Taylor's National Patriotic Front of Liberia (NPFL) and its support for the Revolutionary United Front (RUF) in Sierra Leone. At the time the RUF came into existence, Taylor was Liberia's president, and NPFL's support of the RUF helped it gain traction in its early development and allowed the group to quickly develop a footprint throughout the country.[30] The RUF helped destabilize Sierra Leone, which was Liberia's main competitor, and gave Taylor access to the country's diamond mines. The RUF's ability to assert control over the diamond trade helped undermine Sierra

---

[i] A striking example is a California physician who is known to have provided as much as $4 million to the LTTE over a decade.[29]

Leone's economy, and the sadistic tactics the group used against civilians caused an exodus of refugees and near-total destruction of the country's political structure.

SUPPORTED RESISTANCE MOVEMENTS

One of the first challenges facing states wishing to utilize external support and resistance movements seeking that support is finding a good match. The process requires decisions by both parties. States must choose which resistance movement will best help them reach their objective. Resistance movements must make similar decisions when courted by potential external supporters. Movements must decide whether to accept the support and, in some cases, may have the option to choose between one or more potential supporters. Moreover, resistance movements can opt to actively seek out external supporters rather than passively waiting for offers of support to materialize. This section analyzes and discusses which resistance movements are likely to receive external support from a state. Factors internal to the group, as well as external factors, are explored.

Preferences and Agency Slack

One way to conceptualize how states decide which resistance movement to support is through the principal–agent model.[j] In choosing to delegate the fulfillment of its foreign policy objectives through a third party, states must manage some challenges. The principal actor in this model, the state, is delegating the performance of its goals to an agent, the resistance movement, which may have a different set of objectives. Ideally, the principal would like to fully control the actions of the agent so that it pursues the principal's goals just as if the principal itself were in the theater.[k]

One of the struggles both parties face in this regard is asymmetric information. Clear communication is a problem in most, if not all, complex human undertakings. While both the principal and the agent articulate their objectives to one another, neither party has absolute guarantees that they are being provided with all of the accurate information needed to make good, informed decisions on how to proceed. Each party has good reason to withhold information in order to make

---

[j] An analysis of the motivation of states to provide external support is provided in the "Motivations for External Support" section.

[k] The discussion of the principal–agent model is based on Salehyan, Gleditsch, and Cunningham.[31]

itself a more attractive partner. State actors, for instance, may tell potential partners that their objective is to overthrow or reform a target government but may secretly plan to only disrupt the target government or bleed it of its resources. Likewise, resistance movements may feign objectives similar to those of the state but use the state's resources for their own purposes.

Additionally, principals have little hope of exerting that much control over the actions of their agents, sometimes called "agency slack." Rwanda, for instance, choose to support the CRD, a resistance movement in the Democratic Republic of the Congo headed by Laurent Kabila. The Tutsi-dominated Rwandan government hoped to unseat the Democratic Republic of Congo (DRC) leader, Mobutu, who supported Hutu militias (the *interahamwe*) during and after the 1994 Rwandan genocide. Mobutu allowed the Hutu militias to use eastern Congo as a safe haven while raiding Rwanda from across the border. The CRD was ultimately successful in driving out Mobutu, and Kabila took control of the government himself. However, he then turned on his Rwandan patrons by supporting, not opposing, the *interahamwe* shortly after gaining power.[32] Rwanda failed to secure any benefit from its agent, the CRD, because Kabila sought to secure the legitimacy of his new regime by aligning with the Hutu militias and acted in accordance with his new preferences. Part of the selection process for states is choosing resistance movements that are less likely to act independently of their patrons.

To lessen the potential impact of agency slack and asymmetric information, states often choose to support groups with similar preferences. During the Cold War, preferences were shaped by the ideological divide between the United States and the Soviet Union. Resistance movements typically fell on one end or the other of the left–right spectrum. States must rely on other indicators of preference after the decrease of ideological conflicts in the post-Cold-War era. One of the easiest selection methods states can follow is to support resistance movements that have ethnic or religious ties. Some resistance movements appeal to a specific ethnicity or religious affiliation that spans the national boundaries of the state in which it operates. The LTTE represented ethnic Tamils in Sri Lanka. India, home to a sizable Tamil population, supported the Tamils for a number of years against the Sinhalese regime in Sri Lanka. Likewise, Shia-dominated, theocratic Iran chose Lebanese Hizbollah, a Shia-based movement, to act as its regional buffer against Israel. Resistance movements representing ethnicities or religions with transnational constituencies are more likely to receive state external support than resistance movements that have only local appeal.[33] Supporting a resistance movement with similar preferences lessens the chance that the resistance movement will act against the preferences of its patron.

Internal Characteristics of Resistance Movements

In selecting resistance movements, states also seek those movements that can successfully challenge the target government. A weak resistance movement is likely to offer a poor return on investment for the state supporter. A state supporter invests sizable resources in resistance movements and expects dividends in achieving its foreign policy objectives. A viable resistance movement might be one with good organization and leadership, group cohesion, and a history of successful operations against the state.

Some researchers speculated that movements with strong, central commands exercising a good deal of control over their members were likely to receive support over those movements without such a structure. In the end, their research did not point to this conclusion. The organizational model adopted by insurgencies should be dictated as much by strategy as by choice. Insurgents, such as the Provisional Irish Republican Army (PIRA) in Northern Ireland, adopt flatter, cellular structures to maximize security when the movement is in its infancy or experiencing pressure from counterinsurgent campaigns. At its inception in 1969, PIRA maintained its traditional, hierarchical military structure emphasizing strict command and control. However, as it faced increasing pressure from British authorities in the mid-1970s, it adopted a cellular structure that compartmentalized its operatives from one another. Operatives only had knowledge of the few members of their cells, preventing captured operatives from compromising the entire organization.[1] The foresight of PIRA leadership in aligning structure with strategy helped the insurgents survive an especially penetrating British counterinsurgent campaign. Agile, successful groups such as PIRA evidence the ability to adapt organizational structure to meet strategic needs.

States elect to support moderately strong resistance movements over very weak or very strong ones. The strength of a resistance movement is measured relative to the state it is targeting. Here, strength is measured according to several indicators, including the number of fighters a group can mobilize, its capacity to procure arms, and conventional fighting capacity compared to the target government.[35] When selecting which resistance movements to support, states optimally seek the strongest available insurgents. However, these groups, already well-established

---

[1] For a discussion of insurgent groups' organizational structure, see Bos.[34]

and successful, are less likely to accept constraints on their actions by external supporters in exchange for resources. The strongest groups are able to procure sufficient resources without external support. The weakest groups, meanwhile, are the least attractive option for state supporters because these groups are untried or unsuccessful. Moderately strong groups are in the "Goldilocks Zone," neither so strong that they can turn down help nor so weak that they are not an attractive option for state supporters. Moderately strong groups, therefore, are the resistance movements most likely to attract, and accept, external support.[36]

External Characteristics

Factors external to resistance movements also impact which groups are selected for external support. States choose to support resistance movements in order to achieve specific foreign policy objectives. To be an attractive option, resistance movements must target foreign governments over which state supporters seek some sort of leverage. For this reason, resistance movements operating in states engaged in international rivalries with other states are more likely to receive external support. Here, an international rival is one that another state perceives as threatening enough to be considered an enemy.[37] Not surprisingly, states are more likely to support resistance movements in foreign states perceived as international rivals.[38] International rivals are also likely to be close to home. Most states select resistance movements in nearby, neighboring states.[39] Alongside these factors, states are also likely to select resistance movements as a counter–external support measure. In this instance, the targeted government is receiving support from another government, so that states compete indirectly through third parties, whether the targeted government or the resistance movement.[40]

Much of the current research related to external support for resistance movements focuses on support for insurgent groups already in conflict with targeted governments. The available research derives mostly from political science research on external intervention in civil wars. In terms of phasing, this means that a great deal of what we currently know about external support derives only from cases occurring during the militarization phase. However, in some instances, external support occurs during earlier stages when a resistance movement is nonviolent or has not yet kinetically engaged the target government. Again, in terms of phasing, external support can occur during the latent and/or incipient phases of resistance. The focus on external support in armed conflict overlooks crucial information, especially on research involving the decision-making process potential external supporters use to select potential resistance movements and vice versa. In some cases, external support for the resistance movements may have begun far sooner than a military conflict. Currently, Iran is supporting Hizbollah's assistance to the beleaguered Assad regime in Syria. However, Iran's support for both Assad and Hizbollah fighters in the conflict began decades ago.

MOTIVATIONS FOR EXTERNAL SUPPORT

External actors' motivations for providing support to a resistance movement are highly dependent on their own objectives. As touched on in the "Types of Actors Providing External Support" section, the motivations and forms of support are linked to the identity of the actor. Diaspora motivations, for example, are considerably different from the motivations of an external state. A state may support a group to gain a strategic foothold in the region in order to exploit resources or gain additional territory. Diasporas may support a movement less for strategic reasons and more out of support of ethnic kinship. The goals the external actor hopes to achieve through support of a resistance movement may align with or be in addition to the goals of the movement it supports.[41] This section offers an overview of motivations for external support, with a focus on motivations for state-sponsored support.

Motivations for Non-State-Actor Support

Before focusing on the motivations for state support, it is worth summarizing the general motivations of non-state actors.[m] In general, non-state actors are motivated more by religious and/or ethnic affinity and shared ideology than are state supporters. These reasons may play a role in a state's decision to provide external support but are rarely the only motivators. For non-state actors, the relevance of these motivators is much easier to understand. While a diaspora may be motivated by a shared identity with the supported movement, similarly, refugees are often motivated by a desire to return home or restore stability in the country from which they are displaced. They retain a strong link to the place and its people, perhaps even more so than diasporas because they may not want to settle permanently as immigrant communities but to return to life in their former country.

Refugees and diasporas may believe that military action is the only option for addressing their grievances, although both groups can be subjected to coercion. A resistance movement may capitalize on the

---

[m] This section draws heavily from the work of Byman et al.[42] A more detailed analysis of motivations of non-state actors can be found in that study, particularly in chapters 3–5.

guilt of the diaspora community to garner support, or it may influence the internal politics of the diaspora, which often remains isolated from the larger community in which the group is settled. Refugees may be forced to support a resistance movement in order to ensure their own safety and access to resources. If rebels control a refugee camp and no other group or government agency can provide support, the refugee population may support the rebels out of fear of survival and not because there is a shared identity or shared set of values and objectives.

Other non-state actors, including rebel groups, NGOs, religious groups, and individuals, are often motivated by a shared ideology. Religious organizations will likely support resistance movements that share a commitment to religious doctrine (e.g., the theological support of Lebanese Hizbollah by clerics in Iran). Some individuals may even consider financial support of insurgencies to be an appropriate method of fulfilling charitable religious obligations. However, wealthy individuals who may serve as key benefactors to a resistance movement may be motivated by things other than shared religious beliefs. Individuals may share an ethnic kinship or a common ideology with a group or feel that the group's cause is just, even if there is no other link to the resistance movement through shared place or identity. NGOs rarely purposely support a resistance movement. Rather, their motivations for providing support are often based on their mission to provide supplies and aid to individuals in need, which may require working through a resistance to channel supplies or inadvertently supplying the movement along with the susceptible population that is the focus of their efforts.

Even though support from diasporas and refugees has arguably become more significant to resistance movements in the past twenty years, support from these groups still differs greatly from state support in scope and nature. State motivations may also be less clear, particularly if the strategic objectives of the state do not align with the resistance movement's, but can be achieved indirectly through the provision of external aid. For these reasons, it is important to devote more attention to state-actor motivations for supporting resistance movements.

Motivations for State-Actor Support

States often have strategic ambitions that serve as the primary motivation for supporting resistance movements, even if shared religious or ethnic identities are an underlying factor, which in some instances may serve as a pretense for support even though the real objectives are much more political in nature. Although many objectives for state-sponsored support exist, this paper focuses on five broad categories: regional influence, destabilization of other governments or rivals, regime change, influence within the resistance movement, and internal security.[n] The pursuit of these objectives often complicates a conflict or emerging conflict. States do not necessarily support a group to lessen the duration of or end a conflict.[43] If one of the above objectives is the motivation for support, the state likely does not intend to help the combatants win but rather to help advance an independent agenda; having one side win may not be necessary to achieve that end.

Regional Influence

A state may back a resistance movement in order to increase its regional influence, which may help it gain control over local rivalries or maintain control over larger swaths of territory. An example is Russia's support of various movements in territories that were formerly part of the Soviet Union. The objective was to maintain dominion over as much territory as possible, and the external support has proved effective. States such as Georgia, Tajikistan, and others had insurgencies that benefited from Russian support, and this support made these states more attuned to Russian concerns and perspectives in the region.[44]

Destabilization of Governments or Rivals

Support of a resistance movement may be a way for a state to weaken an enemy state or group without having to engage in direct conflict. Conflict requires a major commitment of resources not to mention exposure to international scrutiny and diplomatic discord. Working through a resistance movement reduces the investment of resources the state must commit and also allows it to plausibly deny involvement if the objective is to keep the support clandestine.

Regime Change

A state may aim to unseat a ruling government and work through a resistance group to achieve this end. This objective can be extremely difficult to achieve because it may require a dramatic increase in resources, especially if the government that it seeks to depose is strong. Moreover, other states may see the destabilization and back groups themselves with the objective of maintaining the current government. In this case, the state seeking to unseat the government will have to overcome a

---

[n] Byman et al. reference thirteen motivating factors, which include those referenced here as well as payback, prestige, support of coreligionists, support of coethnics, irredentism (seeking control of territories governed by another state on the basis of common ethnicity or perhaps prior historical possession), and leftist ideology. As mentioned, although ethnic, religious, and ideological motivations may play a role in state support of a resistance movement, they are rarely the sole or dispositive factor.

particularly formidable set of circumstances and may settle for weakening the rival and attempting to extend its influence in other ways.

Ensuring Influence within the Resistance Movement

In some instances, a state may support a resistance movement in order to ensure some level of control over the group and keep it from adopting policies that are counter to the state's interests. It may also wish to gain influence over the movement so it can later shape or change the group's goals, which will allow the state to exert influence through the resistance movement for future initiatives.

Internal Security

A state may support a resistance movement to counter dissidents within its own territory and maintain security. An insurgency within a state's territory may stage attacks on targets outside of the state. Although the insurgency does not necessarily exist to threaten the state, in order to ensure that the insurgency does not undermine the state's standing with the local population or its objectives, it may support a resistance movement to counter the insurgents' influence and ensure internal security.

FORMS OF EXTERNAL SUPPORT

In its starkest, stylized form, every militarized resistance movement needs three basic things to survive: beds, beans, and bullets. What this list underscores is that a resistance movement seeking to overthrow a government or occupying power absolutely needs a highly motivated force that is adequately resourced and has a haven from which it can operate.

There are a number of ways for a resistance movement to begin to equip itself. It can acquire materials through theft, ambush, local sympathizers, or black-market purchases. Support can also come from nations that are not actively participating in the conflict but that seek to influence its outcome. States can provide all types of support to an insurgent. Some types of support clearly matter more than others. Over time, through the development of UW case studies, this research will identify which types of support matter most and how they are best provided to the resistance movement.

Even though we do not know which types of assistance matter the most, from our initial research for this paper, we have identified ten important forms of support:

1. Safe haven

2. Propaganda
3. Training
4. Weapons and other materials
5. Organizational assistance
6. Financial resources
7. Logistics
8. International political support
9. Direct military support
10. Intelligence

As we begin to develop our knowledge of how states support indigenous resistance movements through the UW case studies, this list will change. However, for the time being, we view the first six types of assistance as being more important than the last four.

No resistance movement can survive for long if it cannot rest, recover, and plan future operations. To this end, every insurgent group needs some type of safe haven. It can be located either in the contested country or outside of it. Although Hizbollah maintained weapons caches and bases in Syria, it also maintained a number of safe havens in Lebanon. By doing so, it kept itself close to the people and tailored its operations to garner support from them. The Palestinian Liberation Organization (PLO) did not always do this, and, as a result, its operations in the West Bank and Gaza Strip suffered and many Palestinians stopped supporting the group.

To be successful, a movement needs to convince people to support, join, and follow it. A resistance group needs a constant stream of new recruits to fight and sustain its operations. Additionally, it must convince the local population that it can govern. As a resistance movement grows, its narrative or raison d'être is critical to its continued success. Oftentimes, narratives are accompanied by sophisticated propaganda campaigns. Presenting the governing authority as feckless and corrupt is essential to convincing a local population that it is better off supporting the resistance. Hizbollah is a master of messaging. It crafts its actions so that they can be scripted into a broader message, namely that Israel persecutes Muslims and that only Hizbollah can keep Lebanon free from Israeli occupation.

By definition, insurgents must operate in a clandestine manner. Additionally, because they are often weaker than the state they oppose, they must resort to guerrilla tactics, which place a premium on surprise and disciplined operational practices. Given this, basic training in tradecraft and fieldcraft is critical. The support that Pakistan's intelligence agency, Directorate for Inter-Service Intelligence (ISI), has provided to Lashkar-e-Taiba has allowed the extremist group to successfully infiltrate and attack targets deep inside India. Furthermore, the ISI's support and training of a number of extremist groups operating in Jammu and Kashmir have, by all accounts, made the groups far more effective and frustrated Indian security forces, leading many observers to consider the fight a draw.

Every resistance movement has materiel needs, to include weapons and explosives. Without access to weapons and other supporting materiel, a movement cannot forcefully challenge an established power and move to the next phase of the conflict. Because of Iran's materiel support, Hizbollah now has one of the most formidable arsenals in the region, and it has successfully employed it against Israel.

Even with the most highly trained and motivated individuals, a resistance movement will always lose to a well-run and disciplined organization. Coordinated effort is critical to challenging an established power. Before Iran's efforts to grow and empower it, Hizbollah was a militant group with a cause and little else. Its ability to have any type of impact, politically or militarily, was at best marginal. With Iran's guidance, which occurred very early in the group's development, Hizbollah built an effective organization capable of disciplined political and military action. Today, Hizbollah is a powerful political party with a potent military arm that by most accounts exceeds the capabilities of the Lebanese military.

For any organization, financial resources are critical to its continued success. Most insurgencies are long. If a resistance movement is to prevail, it must be able to recruit new members, compensate those in the fold, have continued access to materials, develop sources of intelligence, and create and distribute a compelling narrative. All of these tasks require access to capital at some level. Over the years, ISI has provided substantial sums to Lashkar-e-Taiba. These resources bought time and space for the organization to grow. Today, Lashkar-e-Taiba appears to be a self-sustaining entity, with legitimate businesses throughout South Asia.

As mentioned above, the remaining four types of support do not appear to be as critical to a resistance group's success. One such example is intelligence. Because a resistance movement is by definition part of the country, it is often closer to and has better access to individuals with key information about the government's aims and capabilities. Although an outside nation will, more often than not, have intelligence capabilities that the indigenous force does not, these capabilities are often of limited value, especially if the information cannot be widely distributed because its existence would compromise either the group's legitimacy or the sources and methods of the supporting nation. For these reasons, there continues to be an open debate about the role of outside intelligence in an insurgency. A similar kind of logic applies to the other kinds of support as well (i.e., their impact on the outcome of the conflict is not as great as the others on the above list).

Although external support can prove to be critical, it comes with a cost. Once a resistance group welcomes outside support, it runs the risk of losing both legitimacy in the eyes of its supporters and control of the group. A commonly recited theme is that groups that rely too much on outside support are often bound to fail. However, this wisdom is open to question and further research. It appears that aid in the early stages of a conflict—primarily during the clandestine organization or incipient phase—is absolutely critical to a group's success. Understanding both when to provide support (i.e., the timing) and how much support to provide (i.e., the level) are crucial considerations, especially if we are to establish a set of best practices for practitioners. The UW case studies, along with the work seeking to correlate the phases of organizational growth with the stages of a resistance movement, are central to this effort and key for the SOF community.

EFFECTIVENESS OF EXTERNAL SUPPORT

What types of support matter and why they matter are important questions. As discussed in the previous section, timing and levels of support are also crucial to our understanding of outside support to indigenous forces. Although there is some literature that examines the effectiveness of external support, all of these questions largely remain open.

The existing literature[45] indicates that the following types of support are most effective:

1. Safe haven
2. Propaganda
3. Training
4. Weapons and other materials
5. Organizational assistance
6. Financial resources

Several very basic questions are not adequately answered in this literature, namely: What are the measures of effectiveness? Do they generalize from one conflict to another? If so, why? To begin to answer all of these questions, it is necessary to conduct a series of comparative case studies on support to indigenous resistance movements. By using the comparative method, it is possible to search for lessons that generalize but that also have enough context to allow readers to learn to adapt them to the circumstances in which they finds themselves.

The literature as it stands today indicates that the six types of support listed above are important. When they should be provided (i.e., at which phase of a resistance group's growth) and at what level they should be provided are open questions that will have to wait until some of the UW case studies are complete.

POTENTIAL CASE STUDIES AND CASE SELECTION

Appendix A lists the cases considered for selection for further research. This section discusses the guidelines and criteria developed to help prioritize the potential case studies that could be analyzed.

1. **Cases that received significant state support.** Only those resistance movements that received significant support from external-state actors are considered. To mitigate difficulties with determining causality, we also selected cases in which states, and no other actors, such as diasporas or refugees, provided crucial support for the resistance movement.

2. **Successful cases.** For the purposes of this research, we first want to investigate those cases of external support in which the external-state sponsor has met with a measure of success in achieving national objectives. As discussed above, this could be support to a resistance movement that has overthrown a government but also a resistance movement that succeeded in destabilizing or disrupting the targeted regime.

3. **Post-Cold War.** Because of the impact of the changes in the international system, we considered cases in which external support spanned both the Cold-War and post-Cold-War eras, or cases of external support that took place solely in the post-Cold-War era.

4. **Method of warfare.** The method of warfare adopted by resistance movements is a potentially important variable to consider. We considered cases that spanned numerous methods, from irregular to more conventionally fought conflicts.

5. **State capacity.** Regimes that exercise little control over their territory are more prone to experience armed resistance. Weak states facing weak challengers are also a more common conflict dyad than strong states and weak challengers in the post-Cold-War era. We considered cases that involved resistance movements in both weak and (relatively)

stronger states.

The following sets of possible case studies are organized by geographic regions (i.e., Middle East, Asia, Africa, other) and generally appear to meet the selection guidelines and criteria.

Middle East

Iraq (2004–present)

- **Insurgents:** Many Islamist groups, including Al-Mahdi Army, Soldiers of Islam (Ansar al-Islam), Islamic Army of Iraq (IAI), Islamic State in Iraq (ISI), Reformation and Jihad Front (RJF), Army of Ansar Al Sunna (JAAS), and Organization of Jihad's Base in the Country of Two Rivers (TQJBR)
- **External support to insurgents:** Iran (particularly to Shia insurgents), Saudi Arabia and other Arab states (particularly to Sunni insurgents), and various non-state actors (refugees, other revolutionary groups, religious organizations, wealthy individuals)

Israel (1987–present, with emphasis on 2006–present, i.e., Gaza war, Lebanon war, and overall Israel–Iran proxy war)

- **Insurgents:** Fatah/PLO, Islamic Resistance Movement (HAMAS), Palestinian Islamic Jihad (PIJ), and Al-Aqsa Martyrs' Brigades (AMB)
- **External support to insurgents:** Iran and Hizbollah as well as various other non-state actors (refugees, diasporas, other revolutionary groups, religious organizations, wealthy individuals)

Lebanon (1982–2000, i.e., up to the point that Israel was driven from Southern Lebanon)

- **Insurgents:** Hizbollah (party of God)
- **External support to insurgents:** Iran, Syria, and various non-state actors (refugees, other revolutionary groups, religious organizations, wealthy individuals)

Syria (2011–present)

- **Insurgents:** Syrian National Council (SNC) and National Coalition for Syrian Revolutionary and Opposition Forces
- **External support to insurgents:** Saudi Arabia and other Arab states as well as various non-state actors (refugees, other revolutionary groups, religious organizations, wealthy individuals)

Asia

Afghanistan (2001–present)

- **Insurgents:** Taliban and Hezb-i-Islam (Islamic Party terrorist group)
- **External support to insurgents:** elements of Pakistan (e.g., ISI) and various non-state actors (refugees, other revolutionary groups, religious organizations, wealthy individuals)

Afghanistan (1992–1996)

- **Insurgents:** Taliban
- **External support to insurgents:** Pakistan and various non-state actors (refugees, other revolutionary groups, religious organizations, wealthy individuals)

India/Kashmir (1989–present)

- **Insurgents:** Hizb al Mujahideen, Harakat al-Ansar, Jammu and Kashmir Liberation Front (JKLF), Lashkar-e-Taiba, and others (collection of Kashmiri insurgents)
- **External support to insurgents:** Pakistan and various non-state actors (other revolutionary groups, religious organizations, wealthy individuals)

Africa

DRC (1998–2001)

- **Insurgents:** MLC, Congolese Rally for Democracy–Goma (CRD-Goma), and Congolese Rally for Democracy–Liberation Movement (CRD-ML)
- **External support to insurgents:** Ugandan support to MLC and CRD-ML; Rwandan support to CRD-Goma plus various non-state actors (other revolutionary groups)

Sierra Leone (1991–2002, with emphasis on 1991–1997)

- **Insurgents:** RUF and Armed Forces Revolutionary Council (AFRC)
- **External support to insurgents:** Liberia plus various non-state actors (other revolutionary groups)

Uganda (1986–2009, with emphasis on 1994–2002)

- **Insurgents:** Lord's Resistance Army (LRA), Uganda's People's Army (UPA), Holy Spirit Movement (HSM), Alliance for Democratic Forces (ADF), West Nile Bank Front (WNBF), and Uganda National Resistance Front II (UNRF II)
- **External support to insurgents:** Sudan during 1994–2002 plus various non-state actors (other revolutionary groups)

Rhodesia (1960–1979)

- **Insurgents:** Zimbabwe African National Union (ZANU), led by Robert Mugabe, and Zimbabwe African People's Union (ZAPU), led by Joshua Nikomo
- **External support to insurgents:** Soviet Union, as well as competing interests of China

Other (Americas, Europe)

Bosnia-Herzegovina (1992–1995)

- **Insurgents:** Serbian Republic of Bosnia-Herzegovina (Srpska or Bosnian Serbs) and Croatian Republic of Bosnia-Herzegovina (HVO or Bosnian Croats)
- **External support to insurgents:** Serbian support to Srpska; Croatian support to HVO

Croatia (1992–1995)

- **Insurgents:** Serbian Republic of Krajina (Croatian Serb Army) revolt in Krajina
- **External support to insurgents:** Serbia

RECOMMENDED CASES

Case Study 1: Iranian Support for Hizbollah in Lebanon (1982–2000, i.e., up to point that Israel was driven from Southern Lebanon)

The main desire of the insurgent (Hizbollah) and its external-state supporters in the short term was to drive out a foreign power (Israel) from southern Lebanon. Their common long-term goal is to continue a Shia Islamic revolution with Iranian external support and ultimately to drive Israel from the Levant. Before Israel withdrew from Lebanon, the insurgents conducted a significant number of suicide bombings.

They also conducted paramilitary operations, psychological operations, shadow-government activities, and other types of guerrilla warfare. External states such as Iran and Syria have provided extensive and varied support to Hizbollah.

This case fulfills nearly all of the criteria and guidelines discussed above. Iran has provided significant, crucial support to Hizbollah. With the withdrawal of Israel from southern Lebanon (with the exception of Sheba Farms), Iran has achieved a significant national objective. Hizbollah's near victory over Israeli forces in the summer of 2006 further consolidates Iran's success in this case. Iran also began its support in the early 1980s, still firmly in the Cold War. Its support continued after the fall of the Soviet Union, so the case spans both of these important time periods. When Iran began supporting Hizbollah, the nascent movement relied on irregular warfare methods, but in the ensuing decades, the group has also incorporated conventional methods, which were levied against Israel in 2006. Lebanon remains a weak state, which is a factor that is important to explore in Iran's success.

Case Study 2: Pakistani Support for Lashkar-e-Taiba in Kashmir/India (1989–present)

The various insurgents and their external-state supporter (Pakistan) have a main desire to drive out a foreign power (India) from their area (Kashmir). They have been fighting for their independence since 1989 by employing guerrilla warfare, including terrorist attacks. Elements of Pakistan's intelligence and military forces have been supporting the Kashmir rebels, and support for these types of attacks on Indian forces is reportedly part of an overall strategy by elements in Pakistan to use various non-state actors to harass and otherwise occupy Indian security forces (e.g., use of the Lashkar-e-Taiba terrorist group in the 2008 attack on Mumbai).

Pakistani support for the Lashkar-e-Taiba also fulfills a number of guidelines and criteria discussed above, with some important variance. Lashkar-e-Taiba insurgents have received crucial, significant support from Pakistan. As an external-state supporter, Pakistan has met success in imposing a high cost on Indian security forces combating Lashkar-e-Taiba and other Kashmiri-based insurgents. This case differs from Iranian support for Hizbollah with regard to several crucial factors. Lashkar-e-Taiba has been active primarily in the post-Cold-War era. The group has relied primarily on irregular methods, combatting a strong, conventional force in India, a relatively strong state (although the rugged terrain of Kashmir has proven difficult).

Case Study 3: Soviet Support for Insurgency in Rhodesia

Both the Soviet Union and China were active supporters of insurgencies in Africa in the 1970s, and Rhodesia is one example. The Soviet Union mainly provided support to ZAPU, led by Joshua Nikomo, through the use of its special operations forces, front companies, paramilitary forces, intelligence agencies, and clandestine networks. An interesting dynamic of this case is China's competing support to ZANU, led by Robert Mugabe, although at times the Soviet Union also provided support to ZANU.

This case addresses a variety of broader topics that weigh on the Soviet Union's practice of UW, including strategic interests and alliances and how they align with those of the groups it works with; Soviet attitudes to collateral damage and noncombatants; and Soviet risk calculations surrounding the willingness to form tactical alliances with groups who, for ideological or religious reasons, are opposed or only loosely aligned with communism.

Case Study 4: Rwandan and Ugandan Support for Various Insurgents in the DRC (1998–2001)

Rwanda and Uganda intervened in the DRC after they were betrayed by the DRC leader Kabila (who they had supported in the overthrow of Mobutu). At that point, Rwanda and Uganda united in armed support of a new rebel alliance. The resultant insurgents owed their existence and very survival to the two external-state supporters (because they were too weak to fight the Kabila forces on their own). The motives for Rwanda and Uganda varied. The former wanted revenge, to crush the previous Hutu forces responsible for the earlier genocide in Rwanda, to ensure its influence over various opposition groups (particularly near its border), and to plunder some of the Congo's diamond wealth. The latter also wanted revenge, to plunder Congo's mineral wealth, to maintain its influence over its neighbor, and to ensure that anti-Ugandan insurgents could not use the Congo to launch their attacks.

Rwandan and Ugandan support for various resistance movements in the Congo also fulfills the criteria and guidelines discussed above. External-state support was both significant and crucial for the supported resistance movements. In the case of Kabila's CRD, Rwanda helped form the resistance movement in its earliest phases. The CRD was ultimately successful in overthrowing Mobutu, but with important consequences for Rwanda when Kabila turned his back on his Rwandan supporters. The CRD and other resistance movements took place in the post-Cold-War era in a weak state facing a weak challenger relying on irregular methods.

CONCLUDING OBSERVATIONS
AND RECOMMENDATIONS

Much of what we currently know about external support revolves around superpower support for resistance movements during the Cold War. External support was common in this era after the Cuban missile crisis, when the United States and the Soviet Union realized that it was simply too costly to confront each other head on, competing thereafter in the periphery for more power and influence. Although some of these cases are interesting, often the proximate causes of the conflicts and the ideological debate that shaped them were squarely rooted in a period with substantial differences from the contemporary world. Resistance movements often couched their motivations in anticolonial, leftist ideologies. External support, when it did come, came from superpowers struggling in predetermined frameworks and with seemingly limitless pockets. This does not mean that there is no utility in studying these cases, but building a body of knowledge of how external support for resistance movements functions in today's world requires that attention be paid especially to cases in the contemporary, post-Cold-War era. It is clear from the research that more attention should be paid to contemporary cases. There is much work to be done to better understand the use of external support. With a deeper understanding of many of the issues surrounding the use of resistance movements to achieve external-state national objectives, it will be easier for senior decision makers to develop policy options that will allow the United States to better confront this threat. To that end, there are several recommendations for follow-on work.

The first recommendation is to expand the body of work devoted to UW to include not only additional UW case studies to help distill best practices but also the development of a typology for understanding UW. The typology will be used to help clarify the various concepts embedded in the definition and to remove ambiguity in terms. It is a starting point for conducting more rigorous research and can be used to develop different methods of analysis.

The second recommendation is to begin to expand the knowledge base of cases of external support, a starting point for this knowledge base being the table of cases included in Appendix A. This table includes a list of cases, with the following information provided for each: start–end dates; participating parties (external supporters); type of insurgency; type of support provided (at a high level); mechanism of support provided; and outcome of the conflict, if known. A more expansive knowledge base would allow for more quantitative analysis of the variables, which provides a different way to understand the phenomenon of UW. Moreover, quantitative analysis can be extremely useful in presenting the case for UW if the results more concisely articulate the resources required, the associated risk, or even the likelihood of success. Although answers to those questions take time to develop, both the expanded knowledge base and the development of a typology are necessary prerequisites to undertaking this type of analysis.

The third recommendation is to ensure that current UW case studies have clearly formulated research objectives to address the key questions raised throughout this paper. If the goal is to better understand the theory that supports UW doctrine, only research appropriately designed to answer those questions will yield useful insights. The research design of the case studies should be developed to allow for theory development. Along with the research design, to make the studies truly valuable, all new research should depend on fieldwork and primary documents, when possible. Although it may be difficult to travel to some locations, there are plenty of areas in the United States and Europe where there are large diasporas with populations from these regions, which would allow team members to conduct interviews. Additionally, there are a number of former group members in prison. These resources should be exploited for valuable information in order to produce research that draws on the most useful information and provides a new perspective for understanding the complicated dynamics of external support to resistance movements.

APPENDIX. CASES CONSIDERED FOR SELECTION FOR FURTHER RESEARCH

Case (Start–End)	Government	Insurgent	External Supporter (to insurgents)	Type of insurgency	Type of Support	Mechanism of Support	Outcome
Afghan-Northern Alliance War (1996–2001)	Afghanistan	UIFSA (Northern Alliance)	Russia, Iran, Tajikistan, Uzbekistan (US and others in 2001)	Revolution	State	Territory, weapons, materiel, funding	Insurgent victory
Afghan-Soviet War (1978–1992)	Afghanistan	Mujahideen	US, Egypt, Saudi Arabia, Iran, China, Pakistan	Revolution	State	Weapons, materiel, training, funding	Insurgent victory
Afghan-Taliban War (1992–1996)	Afghanistan	Taliban	Saudi Arabia, Bahrain, Qatar, Uzbekistan, Pakistan	Revolution	State	Weapons, materiel, funding	Insurgent victory
Afghan-Taliban War (2001– Ongoing)	Afghanistan	Taliban	Al Qaeda (alleged Iran and Pakistan)	Revolution	Other	Military and intelligence infrastructure (alleged weapons from Iran)	Ongoing
Algerian Independence War (1954–1962)	Algeria	FLN	Morocco, Tunisia	Rebellion	State	Territory	Insurgent victory

213

Case (Start–End)	Government	Insurgent	External Supporter (to insurgents)	Type of insurgency	Type of Support	Mechanism of Support	Outcome
Algerian Islamists War (1992–Ongoing)	Algeria	GIA, FIS, AQIM	Sudan, Iran, IIRO, Algerian diaspora	Revolution	State, diaspora, other	Funding, weapons	Ongoing
Angolan Independence War (1961–1974)	Angola	UNITA, MPLA, FNLA	Soviet Union, China, Cuba, US, Zaire	Rebellion	State	Territory, weapons, materiel, training, funding	Insurgent victory
Angolan War (1975–2002)	Angola	UNITA, FNLA	US, Zaire/DRC, South Africa	Revolution	State	Troops, territory, weapons, materiel, training, funding	Draw
Cabinda Independence War (1991–Ongoing)	Angola	FLEC	Zaire/DRC	Rebellion	State	Territory, weapons, materiel, intelligence	Ongoing
South Yemeni Independence War (1994–1994)	Arab Republic of Yemen	DRY	Saudi Arabia	Rebellion	State	Weapons	Government victory
Nagorno-Karabakh Independence War (1991–1994)	Azerbaijan	NKR	Armenia	Rebellion	State	Troops, weapons, training, funding	Draw
Chittagong Hill Tracts Independence War (1975–1992)	Bangladesh	JSS/SB	India (1975–1977)	Rebellion	State	Access to territory, weapons, materiel, training	Draw
Bosnian-Serb Independence War (1992–1995)	Bosnia	Republica Srpska, Croatian Republic of Herzeg-Bosnia	Croatia, Yugoslavia	Rebellion	State	Territory, weapons, materiel, training, funding	Government victory
Burmese Leftist War (1948–1988)	Burma	Communist Party of Burma	China	Revolution	State	Training, funding	Government victory
Kachin Independence War (1961–1992)	Burma	KIO	China, India	Rebellion	State	Funding	Draw
Karen Independence War (1949–Ongoing)	Burma	KNU	India	Rebellion	State	Funding	Ongoing
Shan Independence War (1959–Ongoing)	Burma	MTA, SSA	Thailand	Rebellion	State	Access to territory	Ongoing
Burundi Hutu War (1991–2008)	Burundi	Palipehutu-FNL, CNDD-FDD	ALIR, FDLR	Revolution	Neighboring co-ethnic rebel group		Draw
Cambodian War (1970–1975)	Cambodia	Khmer Rouge	North Vietnam	Revolution	State	Troops	Insurgent victory
Cambodian War (1978–1998)	Cambodia	Khmer Rouge, KPNLF, FUNCINPEC (Royalists)	China, Thailand, US	Revolution	State	Territory, weapons, materiel, funding	Draw
Cambodian War (1978–1978)	Cambodia (Khmer Rouge)	KNUFNS	Vietnam	Revolution	State	Troops	Insurgent victory
Chad War (1966–1972)	Chad	FROLINAT	Libya, Sudan	Revolution	State	Troops, territory	Draw
Chad War (1976–1982)	Chad	FAN, FAP, FAT	US, France, Sudan, Egypt	Revolution	State	Weapons, training, access to territory (Sudan)	Insurgent victory
Chad War (1991–Ongoing)	Chad	FUCD, RADF, UFDD, AN, UFR	Sudan	Revolution	State	Weapons, materiel, funding	Ongoing

Case (Start–End)	Government	Insurgent	External Supporter (to insurgents)	Type of insurgency	Type of Support	Mechanism of Support	Outcome
Chad War (1983–1990)	Chad (FAN)	GUNT	Libya	Revolution	State	Territory, troops, weapons, funding	Insurgent victory
Tibetan Independence War (1950–1951)	China	Tibetan resistance groups	US	Rebellion	State	Materiel, training	Government victory
Tibetan Independence War (1956–1959)	China	Tibetan resistance groups	US	Rebellion	State	Materiel, training	Government victory
Colombian War (1964–Ongoing)	Colombia	FARC, M-19, ELN, EPL	Venezuela, IRA, ETA (all alleged)	Revolution	State, other insurgents		Ongoing
Congo-Brazzaville War (1997–2002)	Congo and supporting militias	Cobras	France, Chad, Angola, Elf Aquitaine	Revolution	State, corporate	Troops (Chad and Angola), weapons, materiel, funding	Draw
Katanga Independence War (1960–1962)	Congo-Leopoldville	Katanga	Belgium, South Africa, France	Rebellion	State	Troops (South African and Belgian mercenaries), training, logistics, weapons, funding	Government victory
Cote d'Ivoire Military War (2002–2004)	Cote d'Ivoire	MJP, MPCI, MPIGO	Liberia, Burkino Faso	Revolution	State	Access to territory, weapons, materiel, training	Draw
Croatian-Serb Independence War (1992–1995)	Croatia	RSK	Yugoslavia	Rebellion	State	Troops, weapons, materiel, training	Government victory
Turco-Cypriot War (1974–1974)	Cyprus	TMT	Turkey	Rebellion	State	Troops	Draw
Democratic Republic of Congo War (1998–2001)	DRC	MLC, RCD	Uganda, Rwanda	Revolution	State	Troops, weapons, materiel, training, funding	Draw
Democratic Republic of Congo War (2006–2009)	DRC	CNDP	Rwanda	Revolution	State	Access to territory, intelligence, materiel, training	Draw
El Salvador War (1979–1991)	El Salvador	FMLN	Soviet Union, Cuba, Vietnam, Nicaragua	Revolution	State	Weapons, materiel, training, funding	Draw
Eritrean Independence War (1964–1991)	Ethiopia	ELF, EPLF	Egypt, Libya, Saudi Arabia, South Yemen, Sudan, Syria, Iraq, Kuwait, UAE, Eritrean diaspora, TPLF (Ethiopian Insurgents)	Rebellion	State, diaspora, other	Access to territory (Sudan), weapons, materiel, funding	Insurgent victory
Ethiopian War (1974–1991)	Ethiopia	Ethnic, leftist, and royalist groups (TPLF, EPDM, EDU, EPRDF, SALF)	EPLF, ELF, Sudan, Saudi Arabia, US	Revolution (EPDM, EDU, EPRDF), rebellion (TPLF, SALF)	State	Access to territory (Sudan), weapons, materiel, training, funding	Insurgent victory
Ogaden Independence War (1976–Ongoing)	Ethiopia	WSLF, ONLF	Somalia, Eritrea	Rebellion	State	Access to territory, weapons, materiel, training, funding	Ongoing
Oromo Independence War (1977–Ongoing)	Ethiopia	OLF	Sudan, Eritrea, ONLF, EPLF, TPLF, Oromo diaspora	Rebellion	State	Access to territory, weapons, materiel, training, intelligence	Ongoing

Case (Start–End)	Government	Insurgent	External Supporter (to insurgents)	Type of insurgency	Type of Support	Mechanism of Support	Outcome
Cambodian Independence War (1946–1953)	France	Khmer Issarak	Viet Minh	Rebellion	Other	Other	Insurgent victory
French-Indochina Independence War (1946–1954)	France	Viet Minh, Pathet Lao, Khmer Issarak	Soviet Union, China	Rebellion	State	Weapons, materiel, training	Insurgent victory
Abkhazian Independence War (1992–1993)	Georgia	Republic of Abkhazia	CCMP, elements of Soviet military, Krasnodar Krai (Soviet district)	Rebellion	Other	Troops, weapons, materiel, training, funding	Draw
South Ossentian Independence War (1991–1992)	Georgia	Republic of South Ossetia	Russia	Rebellion	State	Troops	Draw
Guinean War (2000–2001)	Guinea	RFDG	RUF	Revolution	Other	Intelligence, infrastructure	Government victory
Guinea-Bissau Military War (1998–1999)	Guinea-Bissau	MJCDPJ	MFDC	Coup d'etat	Other	Intelligence, infrastructure	Insurgent victory
Assamese Independence War (1990–2010)	India	ULFA	NSCN, UNLF, KNU, KIO, PREPAK, NDFB, Myanmar, Bhutan	Rebellion	State, other	Access to territory, weapons, materiel, training, intelligence	Draw
Bodoland Independence War (1989–2004)	India	BDSF, NDFB	Bhutan, Bangladesh, Myanmar, NSCN, ULFA	Rebellion	State, other	Access to territory, training, intelligence	Draw
Indian Naxalite War (1967–Ongoing)	India	Marxist insurgents (PWG, MCC, CPI-M)	LTTE, Pakistan, Nepal, CCOMPOSA, CPN-M	Revolution	State, other	Weapons, materiel, training, intelligence	Ongoing
Kashmiri Independence War (1989–Ongoing)	India	Kashmir Insurgents	Pakistan	Rebellion	State	Access to territory	Ongoing
Manipur Independence War (1982–2009)	India	PLA, UNLF, KCP, PREPAK	KNU, KIO, NSCN, ULFA, Pakistan, Bangladesh	Rebellion	State, other	Access to territory (Bangladesh), weapons, materiel, training, intelligence	Draw
Nagaland Independence War (1956–1997)	India	NSCN	Pakistan, Bangladesh, Myanmar, China, UNLF, KNU, KIO	Rebellion	State, other	Weapons, materiel, training, funding	Draw
Punjabi Sikh Independence War (1982–1993)	India	Sikh Insurgents	Sikh diaspora	Rebellion	Diaspora	Funding	Government victory
Tripura Independence War (1978–2006)	India	TNV, ATTF, NLFT	MNF, Bangladesh, ULFA, NSCN, others	Rebellion	State, other	Weapons, materiel, training, intelligence,	Draw
Iranian Anti-Khomeini War (1979–2001)	Iran	MEK	Iraq	Revolution	State	Access to territory, funding	Government victory
Iranian Kurdish Independence War (1979–1989)	Iran	KDPI	PUK, Iraq	Rebellion	State, other	Weapons, intelligence	Ongoing
Iraqi Islamist War (2004–Ongoing)	Iraq	Al-Mahdi Army, Ansar al-Islam, TQJBR, IAI	Iran, Al Qaeda	Revolution	State, other	Weapons, materiel, training, funding	Ongoing
Iraqi Kurdish Independence War (1961–1996)	Iraq	PUK, PDK	Iran, Syria	Rebellion	State	Troops, weapons, intelligence	Government victory

Case (Start–End)	Government	Insurgent	External Supporter (to insurgents)	Type of insurgency	Type of Support	Mechanism of Support	Outcome
Iraqi Shia War (1985–1996)	Iraq	SCIRI	Iran	Revolution	State	Access to territory, weapons, materiel, training, funding	Draw
First Israeli Intifada (1987–1993)	Israel	PLO, PFLP, Fatah, Hamas	Arab League, Syria, Lebanon, Iran	Rebellion	State, other	Access to territory, training, funding	Draw
Israeli-Gaza War (2006–2009)	Israel	Hamas	Syria, Iran	Rebellion	State	Access to territory, weapons, materiel, funding	Draw
Second Israeli Intifada (2000–2005)	Israel	Palestinian Authority, Fatah, Hamas, PFLP, Islamic Jihad	Syria, Iran	Rebellion	State	Access to territory, training, funding	Draw
Jordanian Palestinian War (1970–1971)	Jordan	PLO	Syria	Revolution	State	Troops	Government victory
Laotian Leftists War (1959–1973)	Laos	Pathet Lao	North Vietnam, China, Soviet Union	Revolution	State	Troops, other	Government victory
Lebanese War (1975–1990)	Lebanon	LNM, LAA, NUF, Amal, Aoun, Hobeika faction	Iraq, Syria, PLO, Lybia	Revolution	State, other	Troops, weapons, materiel, training, intelligence, funding	Government victory
Liberian War (1989–1995)	Liberia	NPFL	Ivory Coast, Burkina Faso, Libya	Revolution	State	Territory, weapons, materiel, training, funding	Insurgent victory
Liberian War (2000–2003)	Liberia	LURD	Guinea	Revolution	State	Territory, weapons, materiel	Insurgent victory
Libyan War (2011–2011)	Libye	NLA	US, Britain, France, Italy and other NATO members	Revolution	State	Troops (air), materiel, funding	Insurgent victory
Azawad Independence War (1990–1994)	Mali	MPA	Libya	Rebellion	State	Weapons	Draw
Mali War (2012–Ongoing)	Mali	MNLA, Ansar Dine	Al Qaeda	Rebellion (MNLA), revolution (Ansar Dine)	Other	Other	Ongoing
Dniestr Independence War (1992–1992)	Moldova	PMR	Elements of Soviet military	Rebellion	Other	Troops, weapons, materiel, training	Draw
Western Sahara Independence War (1975–1989)	Morocco	Polisario	Algeria, Libya	Rebellion	State	Territory, weapons, materiel, training, funding	Government victory
Mozambique War (1977–1992)	Mozambique	Renamo	Rhodesia, South Africa, Kenya, Mozambique diaspora	Revolution	State, diaspora	Territory, weapons, materiel, training, funding	Draw
Nepal Maoist War (1996–2006)	Nepal	CPN-M	Alleged support from MCC, LTTE, KLO, ULFA, CPI-M	Revolution	Other	Weapons, training, intelligence	Draw
Nicaraguan Contra War (1982–1990)	Nicaragua	Contras/FDN	US, Honduras, Saudi Arabia	Counter-revolution	State	Territory (Honduras), weapons, materiel, training, funding	Draw

217

Case (Start–End)	Government	Insurgent	External Supporter (to insurgents)	Type of insurgency	Type of Support	Mechanism of Support	Outcome
Nicaraguan Sandinista War (1978–1979)	Nicaragua	Sandanistas (FSLN)	Cuba	Revolution	State	Weapons, materiel, training	Insurgent victory
Biafra Independence War (1967–1970)	Nigeria	Republic of Biafra	Israel, France, Portugal, South Africa	Rebellion	State	Weapons, materiel, humanitarian supplies	Government victory
Nigerian Islamist War (2009–Ongoing)	Nigeria	Boko Haram	Al Qaeda (allegedly)	Revolution	Other	Unknown	Ongoing
North Yemeni Royalists War (1962–1970)	North Yemen	Royalist Partisans	Saudi Arabia, Jordan, Britain	Coup d'etat, followed by counter-revolution	State	Weapons, materiel	Government victory
Dhofar Independence War (1972–1975)	Oman	PFLO	South Yemen	Rebellion	State	Troops, access to territory	Government victory
Baluchi Independence War (1974–1977)	Pakistan	Baluchi separatists	Afghanistan	Rebellion	State	Territory, weapons, materiel, funding	Government victory
Bengali Independence War (1971–1971)	Pakistan	Bangladesh	India	Rebellion	State	Troops, access to territory, weapons, materiel, funding	Insurgent victory
Pakistani Islamist War (2007–Ongoing)	Pakistan	TNSM, TTP	Taliban, Al Qaeda	Revolution	State, other	Intelligence, infrastructure	Ongoing
Bougainville Independence War (1989–1996)	Papua New Guinea	BRA	Alleged support from Solomon Islands	Rebellion	State (alleged)	Unknown	Draw
Sino-Nationalists War (1949–1958)	People's Republic of China	Republic of China	US	Revolution, rebellion	State	Weapons, materiel, training, funding	Government victory
Moro Independence War (1970–Ongoing)	Philippines	MNLF, MILF, ASG	Libya, Iran, CPP, Jemaah Islamiya, Al Qaeda	Rebellion	State, other	Weapons, materiel, training, funding, intelligence	Ongoing
Philippine Communist War (1969–Ongoing)	Philippines	CPP	MNLF, MILF	Revolution	Other	Weapons, materiel, training, intelligence	Ongoing
Mozambique Independence War (1964–1974)	Portugal	FRELIMO	Soviet Union, China, Cuba, Tanzania, Zambia	Rebellion	State	Weapons, materiel, training	Insurgent victory
Sino-Communist War (1946–1949)	Republic of China	CPC/PLA	Soviet Union	Revolution	State	Weapons, materiel	Insurgent victory
Chechen Independence War (1994–1996)	Russia	Chechen Republic of Ichkeria	Alleged support from Al Qaeda, Taliban and other Islamist organizations	Rebellion	Other	Training, funding	Draw
Rwandan Hutu War (1997–2002)	Rwanda	ALiR	DRC	Revolution	State	Weapons, materiel, training	Draw
Rwandan Hutu War (2009–Ongoing)	Rwanda	FDLR	DRC	Revolution	State	Weapons, materiel	Ongoing
Rwandan Tutsi War (1990–1994)	Rwanda	FPR	Uganda	Revolution	State	Access to territory	Insurgent victory
Casamance Independence War (1990–Ongoing)	Senegal	MFDC	Guinea-Bissau, Libya, Iraq	Rebellion	State	Access to territory, weapons	Ongoing
Sierra Leone War (1991–2002)	Sierra Leone	RUF, AFRC	Burkina Faso, Libya, Liberia	Revolution	Other	Territory, weapons, materiel, training	Government victory

Case (Start–End)	Government	Insurgent	External Supporter (to insurgents)	Type of insurgency	Type of Support	Mechanism of Support	Outcome
Somalia War (1982–1991)	Somalia	SSDF, SNM, SPM, SNM	Ethiopia	Revolution	State	Access to territory	Insurgent victory
Somalia War (1991–1996)	Somalia	USC/SNA	Alleged support from Libya	Revolution	State	Funding	Draw
Somalia War (2006–Ongoing)	Somalia	Al-Shabaab, ARS/UIC, Hizbul Islam, Harakat Ras Kamboni	Eritrea	Revolution	State	Weapons, materiel, training, funding	Ongoing
Inkatha-ANC War (1981–1988)	South Africa	ANC	East Germany, Soviet Union, Cuba, Tanzania, Angola, Mozambique, Zambia	Revolution	State	Territory, weapons, materiel, training, funding	Insurgent victory
Namibian Independence War (1966–1988)	South Africa	SWAPO	UNITA, Soviet Union, Zambia, China, North Korea, OAU Liberation Committee	Rebellion	State, other	Territory, weapons, materiel, training, funding	Insurgent victory
Vietnamese War (1955–1975)	South Vietnam	Viet Cong	North Vietnam, China	Revolution	State	Territory, weapons, materiel, training, funding	Insurgent victory
Tamil Independence War (1984–2009)	Sri Lanka	LTTE	India, Tamil diaspora	Rebellion	State, diaspora	Territory, weapons, training, funding	Government victory
Southern Sudan Independence War (1963–1972)	Sudan	SSLM	Ethiopia, Uganda, Israel	Rebellion	State	Weapons, materiel, training, funding	Draw
Sudanese Darfur War (2003–Ongoing)	Sudan	SLM/A, JEM	Chad, Eritrea, Libya	Rebellion	State	Weapons, materiel, training, funding	Ongoing
Sudanese War (1983–2004)	Sudan	SPLA	Ethiopia, Uganda, Eritrea, Israel, Egypt	Rebellion	State	Territory, weapons, materiel, training, funding	Draw
Syrian War (2011–Ongoing)	Syria	Free Syrian Army, Islamic Front, Al-Nustra, others	Saudi Arabia, Qatar, Turkey, US, Al Qaeda and other Islamist Groups	Revolution	State, other	Weapons, materiel, training, funding	Ongoing
Tajik War (1992–1998)	Tajikistan	UTO	Afghanistan	Revolution	State	Access to territory	Draw
Thai Leftists War (1974–1982)	Thailand	CPT	China, Laos, Vietnam	Revolution	State	Territory, weapons, materiel, training	Government victory
Turkish-Kurdish Independence War (1984–Ongoing)	Turkey	PKK	Syria, Iran, Greece, PJAK, Kurdish diaspora	Rebellion	State, diaspora, other	Territory, weapons, materiel, training, intelligence, funding	Ongoing
Ugandan War (1979–1979)	Uganda	Fronasa, KM, UNLF	Tanzania	Revolution	State	Troops	Insurgent victory
Ugandan War (1980–1986)	Uganda	FUNA, NRA, UNRF	Libya	Revolution	State	Weapons, training	Insurgent victory
Ugandan War (1986–2009)	Uganda	UPA, LRA, ADF	Sudan, Zaire/DRC	Revolution	State	Territory, weapons, materiel, training	Draw
Northern Irish Independence War (1969–1999)	United Kingdom	IRA	Libya, Irish diaspora	Rebellion	State, diaspora	Weapons, funding	Draw
Aden Emergency (1964–1967)	United Kingdom (Aden Protectorate)	NLF, FLOSY	Egypt, North Yemen	Rebellion	State	Territory, weapons	Insurgent victory
Malayan Independence War (1948–1957)	United Kingdom (Federation of Malaya)	MNLA	Soviet Union, China	Rebellion	State	Unknown	Government victory

Case (Start–End)	Government	Insurgent	External Supporter (to insurgents)	Type of insurgency	Type of Support	Mechanism of Support	Outcome
Yemeni Islamist War (2004–Ongoing)	Yemen	AQAP		Revolution			Ongoing
Zaire Shaba War (1977–1978)	Zaire	FNLC	Angola (alleged)	Revolution	State	Unknown	Government victory
Zaire Simba War (1964–1965)	Zaire	PSA	Cuba, Soviet Union	Revolution	State	Troops, weapons, materiel, training, funding	Government victory
Zaire War (1996–1997)	Zaire	AFDL	Uganda, Zambia, Rwanda, Barrick Gold Corporation	Revolution	State	Troops (Uganda, Rwanda), territory, weapons, materiel, training, funding	Insurgent victory

NOTES

[1] US Department of the Army, Joint Publication (JP) 3-05, *Special Operations* (Washington, DC: Joint Chiefs of Staff), July 14, 2014.

[2] Peter Wallensteen and Margareta Sollenberg, "Armed Conflicts, Conflict Termination and Peace Agreements, 1989–96," *Journal of Peace Research* 34, no. 3 (1997): 339–358. Conflict peaked in the early 1990s when there were as many as fifty active conflicts worldwide, defined as organized violence leading to more than twenty-five battle deaths per annum.

[3] Lotta Themnér and Peter Wallensteen, "Armed Conflicts, 1946–2011," *Journal of Peace Research* 49, no. 4 (2012): 565–575.

[4] Joakim Kreutz, "How and When Armed Conflicts End: Introducing the UCDP Conflict Termination Dataset," *Journal of Peace Research* 47, no. 2 (2010): 243–250.

[5] Monica Duffy Toft, *Securing the Peace: The Durable Settlement of Civil Wars* (Princeton, NJ: Princeton University Press, 2010).

[6] Kreutz, "How and When Armed Conflicts End."

[7] Toft, *Securing the Peace: The Durable Settlement of Civil Wars.*.

[8] Ibid..

[9] Paul Collier and Anke Hoeffler, *Greed and Grievance in Civil War*, World Bank Research Working Paper 2355 (Washington DC: World Bank, 2010).

[10] Stathis N. Kalyvas and Laia Balcells, "International System and Technologies of Rebellion," *American Political Science Review* 104, no. 3 (2010): 415–429.

[11] Mary Kaldor, *New and Old Wars: Organized Violence in a Global Era* (Cambridge, England: Polity Press, 2012).

[12] Stathis N. Kalyvas, "The Changing Character of Civil Wars," 2011, Unpublished, 202–219.

[13] Jason Lyall and Isaiah Wilson III, "Rage Against the Machines: Explaining Outcomes in Counterinsurgency Wars," *International Organization* 63 (2009): 67–106.

[14] Kalyvas and Balcells, "International System and Technologies of Rebellion."

[15] Ibid.

[16] Erica Chenoweth and Maria J. Stephan, *Why Civil Resistance Works: The Strategic Logic of Nonviolent Conflict* (New York: Columbia University Press, 2011).

[17] David E. Cunningham, "Blocking Resolution: How External States Can Prolong Civil Wars," *Journal of Peace Research* 43, no. 2 (2010): 115–127.

18. Erin N. Hahn and W. Sam Lauber, *Legal Implications of the Status of Persons in Resistance* (Fort Bragg, NC: United States Army Special Operations Command, forthcoming), chaps. 1, 5, and 6.

19. Daniel Byman et al., *Trends in Outside Support for Insurgent Movements* (Santa Monica, CA: RAND, 2001).

20. Ibid.

21. Ibid., xiv, 71.

22. Idean Salehyan, Kristin Skrede Gleditsch, and David E. Cunningham, "Explaining External Support for Insurgent Groups," *International Organization* 65 (2011): 712–717.

23. Ibid., 709–744.

24. Christopher Paul, "As a Fish Swims in the Sea: Relationships between Factors Contributing to Support for Terrorist or Insurgent Groups Studies," *Studies in Conflict & Terrorism*, 33 (2010): at 492–493.

25. Byman et al., xiv, 1–2.

26. Ibid., 18–23.

27. Ibid., 41.

28. Ibid., 42–43.

29. Ibid., 80.

30. Ibid., 74–76.

31. Salehyan, Gleditsch, and Cunningham, "Explaining External Support for Insurgent Groups," 709–744.

32. "Peace in the Congo: Rwanda and Uganda at Odds," *Strategic Comments* 5, no. 7 (1999): 1–2.

33. Salehyan, Gleditsch, and Cunningham, "Explaining External Support for Insurgent Groups," 709–744.

34. Nathan Bos, Jason Spitaletta, and SORO authors, "Organizational Structure and Function," in *Human Factors Considerations of Undergrounds in Insurgencies,* ed. Nathan Bos (Fort Bragg, NC: United States Army Special Operations Command, 2012), 33–88.

35. David E. Cunningham, Kristin Skrede Gleditsch and Idean Salehyan, "It Takes Two: A Dyadic Analysis of Civil War Duration and Outcome," *Journal of Conflict Resolution* 53, no. 4 (2009): 570–597.

36. Salehyan, Gleditsch, and Cunningham, "Explaining External Support for Insurgent Groups," 709–744.

37. William R. Thompson, "Identifying Rivals and Rivalries in World Politics," *International Studies Quarterly* 45, no. 4 (2001): 557–586.

38. Salehyan, Gleditsch, and Cunningham, "Explaining External Support for Insurgent Groups," 709–744.

39. Byman et al., *Trends in Outside Support for Insurgent Movements.*

40. Salehyan, Gleditsch, and Cunningham, "Explaining External Support for Insurgent Groups," 709–744.

41. Cunningham, "Blocking Resolution: How External States Can Prolong Civil Wars," 115–127.

42. Byman et al., *Trends in Outside Support for Insurgent Movements.*

43. Cunningham, "Blocking Resolution: How External States Can Prolong Civil Wars," 116–117.

44. Byman et al., *Trends in Outside Support for Insurgent Movements.*

45. Daniel Byman, "Outside Support for Insurgent Movements," *Studies in Conflict and Terrorism* 26, no. 12 (2013): 981–1004, and Byman et al. *Trends in Outside Support for Insurgent Movements.*

BIBLIOGRAPHY

Al Jazeera and Agencies. "Gaddafi Loses More Libyan Cities." *Al Jazeera English*, February 24, 2011, http://www.aljazeera.com/news/afr ica/2011/02/2011223125256699145.html.

Balmforth, Richard, and Pavel Polityuk. "Ukraine: Rebels Have Shot down a Ukrainian Military Plane." *Huffington Post*, August 7, 2014, http://www.huffingtonpost.com/2014/08/07/rebels-ukraine-plane_n_5659861.html.

Bariyo, Nicholas. "South Sudan Peace Talks Reach Apparent Break-through." *Wall Street Journal*, June 11, 2014, http://online.wsj.com/articles/south-sudan-peace-talks-reach-apparent-break-through-1402473616.

Barker, Anne. "Time Running Out for Cornered Gaddafi." *ABC News*, February 24, 2011, http://www.abc.net.au/news/2011-02-24/time-running-out-for-cornered-gaddafi/1955842.

BBC News. "Hezbollah Heartlands Recover with Iran's Help." June 12, 2013, http://www.bbc.com/news/world-middle-east-22878198.

——. "How the Taliban Gripped Karachi." March 20, 2013, http://www.bbc.com/news/world-asia-21343397.

——. "Libya: UN backs Action Against Colonel Gaddafi." March 18, 2011, http://www.bbc.co.uk/news/world-africa-12781009.

——. "Libya Conflict: UN Agreement Sought to Unfreeze Assets." August 25, 2011, http://www.bbc.co.uk/news/uk-politics-14661504.

——. "Who are Hamas?" Updated January 4, 2009, http://news.bbc.co.uk/2/hi/1654510.stm.

BBC News Europe. "Ukraine Crisis: Russia Isolated in UN Crimea Vote." March 15, 2014, http://www.bbc.com/news/world-europe-26595776.

Ben Solomon, Ariel. "Hamas Tells Social Media Activists to Always Call the Dead 'Innocent Civilians.'" *Jerusalem Post*, July 21, 2014.

Bergen, Peter L. *Holy War, Inc.: Inside the Secret World of Osama Bin Laden*. New York: The Free Press, 2001.

Berman, Eli. *Radical, Religious, and Right: The New Economics of Terrorism*. Cambridge, MA: Massachusetts Institute of Technology Press, 2009.

Berman, Eli, and David D. Laitin. "Religion, Terrorism and Public Goods: Testing the Club Model." *Journal of Public Economics* 92, nos. 10–11 (2008): 1942–1967.

Bishop, Patrick, and Eamonn Mallie. *The Provisional IRA*. London: Corgi, 1992.

Bolivar, Alberto. "Peru." In *Combating Terrorism: Strategies of Ten Countries*. Edited by Y. Alexander, 84–115. Ann Arbor: University of Michigan Press, 2005.

Boone, Jon. "Taliban Target Mobile Phone Masts to Prevent Tip-offs from Afghan Civilians." *The Guardian*, November 11, 2011.

Bos, Nathan, ed. *Human Factors Considerations of Undergrounds in Insurgencies*. Fort Bragg, NC: United States Army Special Operations Command, 2013.

Bos, Nathan, Jason Spitaletta, and SORO authors. "Organizational Structure and Function." In *Human Factors Considerations of Undergrounds in Insurgencies*, edited by Nathan Bos, 33–88. Fort Bragg, NC: United States Army Special Operations Command, 2012.

Brisard, Jean-Charles. "AQIM Kidnap-for-Ransom Practice." *The Terror Finance Blog* (blog), September 27, 2010, http://www.terrorfinance. org/the_terror_finance_blog/2010/09/aqim-kidnap-for-ransom-practice-a-worrisome-challenge-to-the-war-against-terrorism-financing.html.

Bronner, Ethan. "Parsing Gains of Gaza War." *New York Times*, January 19, 2009.

Buchheit, Frank M. "The Task Force Targeting Meeting: Operations Synchronization." *Infantry Magazine*, May–June 2008.

Bueno de Mesquita, Ethan, and Eric Dickson. "The Propaganda of the Deed: Terrorism, Counterterrorism, and Mobilization." *American Journal of Political Science* 51, no. 2 (2007): 364–381.

Buikema, Ron, and Matt Burger. "Fuerzas Armadas Revolucionarias De Colombia (FARC)." In *Casebook on Insurgency and Revolutionary Warfare, Volume II: 1962–2009*. Edited by Chuck Crossett, 39–70. Fort Bragg, NC: United States Army Special Operations Command, 2009.

Buikema, Ron, and Matt Burger. "New People's Army (NPA)." In *Casebook on Insurgency and Revolutionary Warfare, Volume II: 1962–2009*. Edited by Chuck Crossett, 5–38. Fort Bragg, NC: United States Army Special Operations Command, 2009.

Burnett, Katharine Raley, Christopher Cardona, Jesse Kirkpatrick, Sanaz Mirzaei, and Summer Newton. "Autodefensas Unidas de Colombia (AUC)." In *Case Studies in Insurgency and Revolutionary Warfare—Colombia (1964–2009)*. Edited by Summer Newton, 269–319. Fort Bragg, NC: United States Army Special Operations Command, forthcoming.

Bushnell, David. *The Making of Modern Colombia*. Los Angeles, CA: University of California Press, 1993.

Butler, Katherine. "Why Did Israel Attack Gaza?" *Huffington Post*, posted January 29, 2009 and updated May 25, 2011, http://www.huffington-post.com/2008/12/29/why-did-israel-attack-gaz_n_153987.html.

Byman, Daniel. "Outside Support for Insurgent Movements." *Studies in Conflict and Terrorism* 26, no. 12 (2013): 981–1004

Byman, Daniel, Peter Chalk, Bruce Hoffman, William Rosenau, and David Brannan, *Trends in Outside Support for Insurgent Movements.* Santa Monica, CA: RAND, 2001.

Carnahan, Michael, William Durch, and Scott Gilmore. *Economic Impact of Peacekeeping.* United Nations, Department of Peacekeeping, March 2006.

Cassara, John. *Hide and Seek: Intelligence, Law Enforcement, and the Stalled War on Terrorist Finance.* Washington, DC: Potomac Books, 2006.

Chandrakanthan, A. J. V. "Eelam Tamil Nationalism: An Inside View," in *Sri Lankan Tamil Nationalism: Its Origins and Development in the Nineteenth and Twentieth Centuries.* Edited by A. Jeyaratnam Wilson. Vancouver: University of British Columbia Press, 2000.

Chenoweth, Erica, and Maria J. Stephan. *Why Civil Resistance Works: The Strategic Logic of Nonviolent Conflict.* New York: Columbia University Press, 2011.

Chernick, Mark. "PCP-SL: Partido Communista de Peru-Sendero Luminoso," in *Terror, Insurgency, and the State: Ending Protracted Conflicts.* Edited by Marianne Heiberg, Brendan O'Leary, and John Tirman. Philadelphia: University of Pennsylvania Press, 2007.

CNN. "Beirut Marine Barracks Bombing Fast Facts." June 13, 2013, www.cnn.com/2013/06/13/world/meast/beirut-marine-barracks-bombing-fast-facts.

Cobb, Charles E. Jr., and Roberto Caputo. "Eritrea Wins the Peace," *National Geographic* 189, no. 6 (1996): 82–106.

Collier, Paul, and Anke Hoeffler. "Greed and Grievance in Civil War." *Oxford Economic Papers* 56, no. 4 (2004): 563–595.

Collier, Paul, and Anke Hoeffler. *Greed and Grievance in Civil War.* World Bank Research Working Paper 2355. Washington DC: World Bank, 2010.

Collier, Paul, Anke Hoeffler, and Dominic Rohner. "Beyond Greed and Grievance: Feasibility and Civil War," *Oxford Economic Papers* 61, no. 1 (2009): 1–27.

Collins, Laura. " 'Jihad Cool': The Young Americans Lured to Fight for ISIS Militants with Rap Videos, Adventurism, and Firsthand Account of the 'Fun' of Guerrilla War." *Daily Mail Online,* June 19, 2014.

Condit, D. M. *Case Study in Guerrilla War: Greece during World War II.* Washington, DC: Special Operations Research Office, American University, 1961.

Conley, Jerome. "Revolutionary United Front (RUF)—Sierra Leone." In *Casebook on Insurgency and Revolutionary Warfare, Volume II: 1962–2009*. Edited by Chuck Crossett, 763–800. Fort Bragg, NC: United States Army Special Operations Command, 2009.

Crenshaw, Martha. "The Causes of Terrorism." *Comparative Politics*, 13 no. 4 (July 1981): 379–399.

Crossett, Chuck, and Summer Newton. "The Provisional Irish Republican Army: 1969–1998," in *Casebook on Insurgency and Revolutionary Warfare, Volume II: 1962–2009*. Edited by Chuck Crossett, 379–421. Fort Bragg, NC: United States Army Special Operations Command, 2009.

Cunningham, David E. "Blocking Resolution: How External States Can Prolong Civil Wars." *Journal of Peace Research* 43, no. 2 (2010): 115–127.

Cunningham, David E., Kristin Skrede Gleditsch and Idean Salehyan. "It Takes Two: A Dyadic Analysis of Civil War Duration and Outcome." *Journal of Conflict Resolution* 53, no. 4 (2009): 570–597.

Davis, Paul K., and Angela O'Mahoney. *A Computational Model of Public Support for Insurgency and Terrorism: A Prototype for More-General Social-Science Modeling*. Santa Monica, CA: RAND Corporation, 2013.

de Carbonnel, Alissa. "RPT-INSIGHT—How the Separatists Delivered Crimea to Moscow." *Reuters*, March 13, 2014, http://in.reuters.com/article/2014/03/13/ukraine-crisis-russia-aksyonov-idINL6N0M93AH20140313.

de la Calle, Luis. "The Repertoire of Insurgent Violence," research paper, prepared for the American Political Science Association Conference, Seattle, September 2001.

de la Calle, Luis, and Ignacio Sanchez-Cuenca. "Rebels without a Territory: An Analysis of Nonterritorial Conflicts in the World, 1970–1997." *Journal of Conflict Resolution* 56, no. 4 (2012): 580–603.

DeVore, Marc R. "Exploring the Iran-Hezbollah Relationship: A Case Study of How State Sponsorship Affects Terrorist Group Decision-Making." *Perspectives on Terrorism* 6, no. 4–5 (October 2012): 85–107.

Dzinesa, Gwinyayi, and Paulo Wache. "Mozambican Elections: What to Make of Dhlakama's Intention to Run for President." *All Africa*, May 27, 2014, http://allafrica.com/stories/201405270478.html.

Ehrenfeld, Rachel. "The Muslim Brotherhood New International Economic Order." *The Terror Finance Blog* (blog), October 13, 2007, http://www.terrorfinance.org/the_terror_finance_blog/2007/10/the-muslim-brot-1.html.

———. "Drug Trafficking, Kidnapping Fund Al Qaeda." *CNN*, May 4, 2011.

El-Qorchi, Mohammed. "The Hawala System." *Finance and Development* 39, no. 4 (December 2002), http://www.gdrc.org/icm/hawala.html.

Erdbrink, Thomas. "With War at its Doorstep, Iran Sees Its Revolutionary Guards in a Kinder Light." *New York Times*, July 18, 2014.

Field Manual 3-05.130 (FM 3-05.130), *Unconventional Warfare*. Washington, DC: Headquarters, Department of the Army, September 2008.

Field Manual 3-24 (FM 3-24), *Counterinsurgency*. Washington, DC: Headquarters, Department of the Army, 2006.

Financial Action Task Force/Groupe d'action financiere. "Terrorist Financing." Paris: OECD, 2008. http://www.fatf-gafi.org/dataoecd/28/43/40285899.pdf.

Flanigan, Shawn Teresa. "Charity as Resistance: Connections between Charity, Contentious Politics, and Terror." *Studies in Conflict & Terrorism* 29, no. 7 (2006): 641–655.

———. "Nonprofit Service Provision by Insurgent Organizations: the Cases of Hizballah and the Tamil Tigers." *Studies in Conflict & Terrorism* 31, no. 6 (2008): 499–519.

Flanigan, Shawn Teresa, and Mounah Abdel-Samad. "Hezbollah's Social Jihad: Nonprofits as Resistance Organizations." *Middle East Policy* 16, no. 2 (June 2009): 122–137.

Fleet Marine Force Reference Publication 12-18, *Mao Tse-tung on Guerilla Warfare*. Translated by Samuel B. Griffith. Washington, DC: Headquarters, Department of the Navy, 1989).

Foreign Staff. "RENAMO suspends Mozambique Ceasefire, Vows to Step Up Attacks." *Business Day Live*, June 6, 2014, http://www.bdlive.co.za/africa/africannews/2014/06/06/renamo-suspends-mozambique-ceasefire-vows-to-step-up-attacks.

"Founding Statement of the Interim Transnational National Council." Lauterpacht Centre for International Law, March 5, 2011, http://www.lcil.cam.ac.uk/sites/default/files/LCIL/documents/arab-spring/libya/Libya_12_Founding_Statement_TNC.pdf.

Franck, Thomas. *The Power of Legitimacy Among Nations*. New York: Oxford University Press, 1990.

Galula, David. *Counterinsurgency Warfare: Theory and Practice*. Westport, CT: Praeger Security International, 2006.

Gervais, Bryan. "Hutu–Tutsi Genocides." In *Casebook on Insurgency and Revolutionary Warfare, Volume II: 1962–2009*. Edited by Chuck Crossett, 307–342. Fort Bragg, NC: United States Army Special Operations Command, 2009.

Giorgis, Andebrhan Welde. *Eritrea at a Crossroads: A Narrative of Triumph, Betrayal, and Hope.* Houston: Strategic Book Publishing and Rights, 2014.

Global Terrorism Database, http://www.start.umd.edu/gtd/.

Goodman, Marc. "How Technology Makes Us Vulnerable." *CNN,* July 29, 2012.

Grange, David, and J. T. Patten. "Assessing and Targeting Illicit Funding in Conflict Ecosystems: Irregular Warfare Correlations." *Small Wars Journal* (September 2009).

Guelke, Adrian, and Jim Smyth. "The Ballot Bomb: Terrorism and the Electoral Process in Northern Ireland." In *Political Parties and Terrorist Groups.* Edited by Leonard Weinberg, 103–124. London: Frank Cass, 1992.

Gurbey, Gulistan. "The Kurdish Nationalist Movement in Turkey Since the 1980s," in *The Kurdish National Movement in the 1990s.* Edited by Robert Olson. Lexington, KY: The University Press of Kentucky, 2006.

Gurr, Ted Robert. *Why Men Rebel.* Princeton, NJ: Princeton University Press, 1970.

Hahn, Erin N., and W. Sam Lauber. *Legal Implications of the Status of Persons in Resistance.* Washington, DC: US Government Printing Office, *forthcoming,* 2014.

Hamzeh, Ahmad Nizar. *In the Path of Hizbullah.* Syracuse: Syracuse University Press, 2004.

Harik, Judith Palmer. *Hezbollah: The Changing Face of Terrorism.* London: I. B. Tauris, 2004.

———. "Hizballah's Public and Social Services and Iran." In *Distant Relations: Iran and Lebanon in the Last 500 Years.* Edited by H. E. Chehabi. London: I. B. Taurus & Co., 2006.

Harkov, Lahav. "Journalists Threatened by Hamas for Reporting Use of Human Shields." *Jerusalem Post,* July 31, 2014.

Hasenfeld, Y. "Power in Social Work Practice." *Social Service Review* 61, no. 3 (1987): 469–483.

Hegghammer, Thomas. "Terrorist Recruitment and Radicalization in Saudi Arabia." *Middle East Policy,* 8, no. 4 (2006): 39–60.

Hoffman, Bruce. *Inside Terrorism.* New York: Columbia University Press, 1998.

Hristov, Jasmin. *Blood and Capital: The Paramilitarization of Colombia.* Athens, OH: Ohio University Press, 2009.

Human Rights Watch. "Iran: Halt the Crackdown: End Violent Attacks on Protesters, Arrests of Critics." June 19, 2009.

———. "World Report 2013: Eritrea." January 1, 2013, http://www.hrw.org/world-report/2013/country-chapters/eritrea.

———. "World Report 2014: Libya." http://www.hrw.org/world-report/2014/country-chapters/libya, 1.

Human Security Centre. *Human Security Report 2005: War and Peace in the 21st Century.* New York: Oxford University Press, 2005.

Huntington, Samuel P. "The Clash of Civilizations?" *Foreign Affairs* 72, no. 3 (1993): 22–49.

IRIN. "Lebanon: The Many Hands and Faces of Hezbollah." March 29, 2006, http://www.irinnews.org/report/26242/lebanon-the-many-hands-and-faces-of-hezbollah.

———. "Sudan: Southern Pull-out Threatens Peace Deal." October 11, 2007, http://www.irinnews.org/report/74746/sudan-southern-pull-out-threatens-peace-deal.

"ISAF Discusses Insurgent Propaganda Messaging." International Security Assistance Force – Afghanistan Headquarters press release, August 5, 2010, http://www.isaf.nato.int/article/isaf-releases/isaf-discusses-insurgent-propaganda-messaging.html.

Islamic Republic of Afghanistan: 2005 Article IV Consultation and Sixth Review under the Staff Monitored Program—Staff Report; Staff Statement; Public Information Notice on the Executive Board Discussion; and Statement by the Executive Director for the Islamic Republic of Afghanistan, Country Report No. 06/113. Washington, DC: International Monetary Fund, March 2006.

Jaber, Hala. *Hezbollah: Born with a Vengeance.* New York: Columbia University Press, 1997.

Jackson, Robert H. *The Global Covenant: Human Conduct in a World of States.* Oxford: Oxford University Press, 2000.

Jaeger, David A., Esteban F. Klor, Sami H. Miaari, and M. Daniele Paserman. "Can Militants Use Violence to Win Public Support? Evidence from the Second Intifada." *Journal of Conflict Resolution* 59, no. 3 (2015): 528–549.

Jawad, Rana. "Libyan Voters Prepare for Change." *BBC News,* July 5, 2012, http://www.bbc.com/news/world-africa-18721576.

Jenkins, B. M. "International Terrorism: A New Mode of Conflict." In *International Terrorism and World Security.* Edited by D. Carlton and C. Schaerf, 13–49. London: Croom Helm, 1975.

Joint Publication 1-02 (JP 1-02), *Department of Defense Dictionary of Military and Associated Terms.* Washington, DC: Department of Defense, November 8, 2010, as amended through August 15, 2014.

Joint Publication (JP) 3-05, *Special Operations*. Washington, DC: Joint Chiefs of Staff, July 14, 2014.

Jones, James W. *Blood That Cries out from the Earth: The Psychology of Religious Terrorism*. Oxford: Oxford University Press, 2008.

Jones, Seth. "Resurgence of al Qaeda." *RAND Review* 36, no. 2 (Fall 2012).

Kaldor, Mary. *New and Old Wars: Organized Violence in a Global Era*. Cambridge, England: Polity Press, 2012.

Kalyvas, Stathis N. "The Changing Character of Civil Wars." 2011, Unpublished.

———. *The Logic of Violence in Civil War*. New York: Cambridge University Press, 2006.

Kalyvas, Stathis N., and Laia Balcells. "International System and Technologies of Rebellion." *American Political Science Review* 104, no. 3 (2010): 415–429.

Kaplan, Robert D. *Balkan Ghosts: A Journey through History*. New York: St. Martin's, 1993.

Kapstein, Ethan, and Kamna Kathuria. *Economic Assistance in Conflict Zones: Lessons from Afghanistan*. Center for Global Development Policy Paper 013, October 2012.

Kasfir, Nelson. "Guerrillas and Civilian Participation: the National Resistance Army in Uganda, 1981–86." *Journal of Modern African Studies* 43, no. 2 (2005): 271–296.

Katzman, Kenneth. *Afghanistan: Post-Taliban Governance, Security, and U.S. Policy*. CRS Report no. RL30588. Washington, DC: Congressional Research Service, July 11, 2014).

Kirkpatrick, Jesse, and Mary Kate Schneider. "I³M–Interest, Identification, Indoctrination, and Mobilization: A Short Introduction to a New Model of Insurgent Involvement." *Special Warfare Magazine* 26, no. 4 (October–December 2013): 23–27.

Kreutz, Joakim. "How and When Armed Conflicts End: Introducing the UCDP Conflict Termination Dataset." *Journal of Peace Research* 47, no. 2 (2010): 243–250.

Kydd, Andrew H., and Barbara F. Walter. "The Strategies of Terrorism." *International Security* 31, no. 1 (2006): 49–80.

Laub, Zachary. "Backgrounder: The Taliban in Afghanistan." Council on Foreign Relations. Updated July 4, 2014, http://www.cfr.org/afghanistan/taliban-afghanistan/p10551.

Laud, Karin, and Josef Federman. "Israel Calls for North Gaza Evacuation after Raid." *Huffington Post*, posted July 13, 2014 and updated September 11, 2014, http://www.huffingtonpost.com/2014/07/13/israel-gaza-evacuation-raid_n_5581540.html.

Law, Randall. *Terrorism: A History*. Cambridge, UK: Polity Press, 2009.

Leonhard, Robert. "Recruiting." In *Undergrounds in Insurgent, Revolutionary, and Resistance Warfare*. Edited by Robert Leonhard, 19–42. Fort Bragg, NC: United States Special Operations Command, 2013.

Levitt, Matthew. "Hamas from Cradle to Grave," *Middle East Quarterly* 11, no. 1 (2004): 1–12.

———. *Hezbollah: The Global Footprint of the Party of God*. Washington, DC: Georgetown University Press, 2013.

Levitt, Matthew, and Dennis Ross. *Hamas: Politics, Charity, and Terrorism in the Service of Jihad*. The Washington Near East Institute, 2006.

Lichbach, Mark. *The Rebel's Dilemma*. Michigan: University of Michigan Press, 1995.

Looney, Robert E. "The Business of Insurgency: the Expansion of Iraq's Shadow Economy," *The National Interest* (Fall 2005): 70.

Lyall, Jason, and Isaiah Wilson III. "Rage Against the Machines: Explaining Outcomes in Counterinsurgency Wars." *International Organization* 63 (2009): 67–106.

Mamdani, Mahmood. *When Victims Become Killers*. Princeton, NJ: Princeton University Press, 2001.

Mampilly, Zachariah Cherian. *Rebel Rulers: Insurgent Governance and Civilian Life during War*. Ithaca, NY: Cornell University Press, 2011.

Manning, Carrie. "Armed Opposition Groups into Political Parties: Comparing Bosnia, Kosovo, and Mozambique." *Studies in Comparative International Development* 39, no. 1 (2004): 54–76.

———. *The Making of Democrats: Elections and Party Development in Postwar Bosnia, El Salvador, and Mozambique*. New York: Palgrave Macmillan, 2008.

Marshall, Shana. "Hizbollah: 1982–2009." In *Casebook on Insurgency and Revolutionary Warfare, Volume II: 1962–2009*. Edited by Chuck Crossett, 552. Fort Bragg, NC: United States Army Special Operations Command, 2009.

Marten, Kimberly. "Why Sanctions against Russia Might Backfire." *Huffington Post*, August 21, 2014, http://www.huffingtonpost.com/kimberly-marten/why-sanctions-against-rus_b_5696038.html.

Masters, Jonathan, and Zachary Laub. "Hezbollah (a.k.a. Hizbollah, Hizbu'llah)." Council on Foreign Relations, January 3, 2014, http://www.cfr.org/lebanon/hezbollah-k-hizbollah-hizbullah/p9155.

Mazen, Maram. "Bloodshed Corrodes Support for Boko Haram." *Al Jazeera*, May 25, 2014.

McCauley, Clark, and Sophia Moskalenko. *Friction: How Radicalization Happens to Them and Us*. Oxford: Oxford University Press, 2011.

———. "Mechanisms of Political Radicalization: Pathways toward Terrorism." Terrorism and Political Violence, 20, no. 3 (2008): 415–433

McCormick, Gordon H., and Frank Giordano. "Things Come Together: Symbolic Violence and Guerrilla Mobilisation." *Third World Quarterly* 28, no. 2 (2007): 295–320.

McGreal, Chris. "U.S. Supreme Court: Nonviolent Aid to Banned Groups Tantamount to 'Terrorism.'" *The Guardian*, June 21, 2010.

Menapolis. "Local Councils in Syria: A Sovereignty Crisis in Liberated Areas." Policy paper. Istanbul: Menapolis, September 2013.

Metelits, Claire. *Inside Insurgency: Violence, Civilians, and Revolutionary Group Behavior.* New York: New York University Press, 2010.

Milgram, Stanley. *Obedience to Authority.* Harper Perennial Modern Classics, 2009.

Mirzaei, Sanaz. "Taliban: 1994–2009." In *Casebook on Insurgency and Revolutionary Warfare, Volume II: 1962–2009.* Edited by Chuck Crossett, 651–684. Fort Bragg, NC: United States Army Special Operations Command, 2009.

Mishal, Shaul, and Avraham Sela. *The Hamas Wind: Violence and Compromise.* Tel Aviv: Miskal, 2006.

Mitchell, George. *Making Peace.* New York: Alfred Knopf, 1999.

Molnar, Andrew R. *Human Factors Considerations of Undergrounds in Insurgencies.* Washington, DC: Special Operations Research Office, American University, 1966.

———. *Undergrounds in Insurgent, Revolutionary, and Resistance Warfare.* Washington, DC: Special Operations Research Office, American University, 1963.

Moloney, Ed. *A Secret History of the IRA.* New York: W. W. Norton, 2002.

Murshed, Syed Mansoob, and Sara Pavan. *Identity and Islamic Radicalization in Western Europe: Economics of Security Working Paper 14.* Berlin: Economics of Security, 2009.

Newton, Summer, and Robert Leonhard. "Shadow Government." In *Undergrounds in Insurgent, Revolutionary, and Resistance Warfare.* Edited by Robert Leonhard, 133–134. Fort Bragg, NC: United States Army Special Operations Command, 2013.

Nix, Maegen, and Shana Marshall, "Liberation Tigers of Tamil Eelam (LTTE)," in *Casebook on Insurgency and Revolutionary Warfare, Volume II: 1962–2009.* Edited by Chuck Crossett, 233–275. Fort Bragg, NC: United States Army Special Operations Command, 2009.

Noonan, Michael. "Ahern Says His View of Sinn Fein Remains the Same." *Irish Times*, May 3, 2001.

Olson, Mancur. *The Logic of Collective Action: Public Goods and the Theory of Groups.* Cambridge, MA: Harvard University Press, 2009.

Parkinson, Sarah Elizabeth. "Organizing Rebellion: Rethinking High-Risk Mobilization and Social Networks in War." *American Political Science Review* 107, no. 3 (2013): 418–432.

Passas, Nikos. "Fighting Terror with Error: The Counter-productive Regulation of Informal Value Transfers." *Crime, Law and Social Change* 45, no. 4–5 (2006): 315–336.

Patel, Faiza. *Rethinking Radicalization.* Brennan Center for Justice, New York University School of Law, 2011.

Paul, Christopher. "As a Fish Swims in the Sea: Relationships between Factors Contributing to Support for Terrorist or Insurgent Groups Studies," *Studies in Conflict & Terrorism,* 33 (2010): at 492–493.

"Peace in the Congo: Rwanda and Uganda at Odds." *Strategic Comments* 5, no. 7 (1999): 1–2.

Pedahzur, Ami, and Leonard Weinberg. *Political Parties and Terrorist Groups.* New York: Routledge, 2013.

Perdoma, María Eugenia Vásquez. *My Life as a Colombian Revolutionary: Reflections of a Former Guerillera.* Philadelphia: Temple University Press, 2005.

Peters, Gretchen. *Seeds of Terror: How Drugs, Thugs, and Crime Are Reshaping the Afghan War.* New York: St. Martin's Press, 2009.

Philipson, Liz, and Yuvi Thangarajah. *The Politics of the North-East: Part of the Sri Lanka Strategic Conflict Assessment 2005.* Colombo, Sri Lanka: The Asia Foundation, 2005.

Pool, David. *From Guerillas to Government: Eritrean People's Liberation Front.* Athens, OH: Ohio University Press, 2001.

Radio Free Europe/Radio Liberty. "Middle East: Why Did Hamas Win Palestinian Poll?" January 26, 2006, http://www.rferl.org/content/article/1065113.html.

Rajan, Raghuram G., and Arvind Subramanian. "Aid, Dutch Disease, and Manufacturing Growth," *Journal of Development Economics* 94, no. 1 (2011): 106–118.

Rands, Richard. "In Need of Review: SPLA Transformation in 2006-2010 and Beyond." *Small Arms Survey.* HSBA Working Paper 23, November 1, 2010, http://www.smallarmssurveysudan.org/fileadmin/docs/working-papers/HSBA-WP-23-SPLA-Transformation-2006-10-and-Beyond.pdf.

Ranstorp, Magnus. "Hezbollah's Command Leadership: Its Structure, Decision-making and Relationship with Iranian Clergy and Institutions." *Terrorism and Political Violence* 6, no. 3 (1994): 303–339.

Rashid, Ahmed. *Taliban: Militant Islam, Oil, and the New Great Game in Central Asia*. London: I. B. Tauris, 2000.

Raychaudhuri, Sumana. "Will Sri Lanka Drive the Tigers to Extinction?" *The Nation*, February 6, 2009, http://www.thenation.com/article/will-sri-lanka-drive-tigers-extinction.

Reguler, Arnon. "Achievement to Hamas in Local Elections: Won Most of Big Municipalities." *Haaretz*, May 8, 2005.

Reich, Walter. *Origins of Terrorism: Psychologies, Ideologies, Theologies, States of Mind*. Washington, DC: Woodrow Wilson Center, 1990.

Reuters. "Mozambique Army Launches Retaliatory Attack on RENAMO." *Yahoo Finance UK & Ireland*, October 18, 2013, https://uk.finance.yahoo.com/news/mozambique-army-launches-retaliatory-attack-183944138.html.

Richard, Norton Augustus. *Hezbollah: A Short History*. Princeton, NJ: Princeton University Press, 2007.

Richards, A. "Terrorist Groups and Political Fronts: the IRA, Sinn Fein, the Peace Process and Democracy." *Terrorism and Political Violence* 13, no. 4 (2001): 72–89.

Robins, Philip. "The Overlord State: Turkish Policy and the Kurdish Issue." *International Affairs* 69, no. 4 (1993): 657–676.

Rochlin, James F. *Vanguard Revolutionaries in Latin America: Peru, Colombia, Mexico*. Boulder, CO: Lynne Rienner Publishers, 2003.

Rodewig, Cheryl. "Geotagging Poses Security Risks." US Army press release, March 7, 2012.

Romero, Mauricio. "Changing Identities and Contested Settings: Regional Elites and the Paramilitaries in Colombia." *International Journal of Politics, Culture and Society* 14, no. 1 (2000): 51–69.

Rone, Jemera, John Prendergast, and Karen Sorensen. *Civilian Devastation: Abuses by All Parties in the War in Southern Sudan*. New York: Human Rights Watch, 1994.

Rosenberg, Matthew. "Breaking with the West, Afghan Leader Supports Russia's Annexation of Crimea." *New York Times*, March 23, 2014, http://www.nytimes.com/2014/03/24/world/asia/breaking-with-the-west-afghan-leader-supports-russias-annexation-of-crimea.html?ref=asia&_r=0.

Rosenzweig, David. "Judge Rules Against Patriot Act." *Los Angeles Times*, January 27, 2004.

Rudoren, Jodi, and Ben Hubbard. "Despite Gains, Hamas Sees a Fight for its Existence and Presses Ahead." *New York Times*, July 27, 2014.

Sageman, Marc. *Understanding Terrorist Networks*. Philadelphia: University of Pennsylvania Press, 2004.

Salehyan, Idean, Kristin Skrede Gleditsch, and David E. Cunningham. "Explaining External Support for Insurgent Groups." *International Organization* 65 (2011): 709–744.

Scheuer, Michael. *Through Our Enemies' Eyes: Osama Bin Laden, Radical Islam, and the Future of America.* Washington, DC: Potomac Books, Inc., 2008.

Schroeder, Rob, Sean Everton, and Russell Shepher. "Mining Twitter Data from the Arab Spring." *CTX2* no. 4 (November 2012).

Scobbie, Iain. "South Lebanon 2006," in *International Law and the Classification of Conflicts.* Edited by Elizabeth Wilmshurst. Oxford: Oxford University Press, 2012.

Scott, James M. *Deciding to Intervene: The Reagan Doctrine and American Foreign Policy.* Durham, NC: Duke University Press, 1996.

Sela, Avraham. *Non-State Peace Spoilers and the Middle East Peace Efforts.* Jerusalem: The Floersheimer Institute for Policy Studies, 2005.

Sengupta, Somini. "Vote by U.N. General Assembly Isolates Russia." *New York Times,* March 27, 2014, http://www.nytimes.com/2014/03/28/world/europe/General-Assembly-Vote-on-Crimea.html.

Shapira, Shimon. *Hezbollah between Iran and Lebanon.* Tel Aviv: Hakibbutz Hameuchad, 2000.

Shuaib, Ali. "Libyan Assembly Votes Gaddafi Opponent as President." *Reuters,* August 9, 2012, http://www.reuters.com/article/2012/08/09/us-libya-assembly-idUSBRE8781ID20120809.

Storm, Darlene. "Intelligence Agencies Hunting for Terrorists in World of Warcraft." *ComputerWorld* (blog), April 13, 2011, http://blogs.computerworld.com/18131/intelligence_agencies_hunting_for_terrorists_in_world_of_warcraft; "Al-Shabab Bans Mobile Phone Money Transfers in Somalia," *BBC,* October 18, 2010, http://www.bbc.co.uk/news/world-africa-11566247.

Tapper, Jake. "Iraqi Insurgents Increasingly Use Internet as Propaganda Machine." *ABC News,* February 14, 2006.

Tekle, Tesfa-Alem. "Eritrean Leader Pledges to Draft New Constitution." *Sudan Tribune,* May 24, 2014, http://www.sudantribune.com/spip.php?article51115.

"Terrorist Organization Profile: Mozambique National Resistance Movement," National Consortium for the Study of Terrorism and Responses to Terrorism, http://www.start.umd.edu/tops/terrorist_organization_profile.asp?id=314.

The Economist. "Why Hamas Fires Those Rockets." July 19, 2014.

The Middle East Research and Information Project. Narges Bajoghli, "Iranian Cyber-Struggles." May 3, 2012.

The News International. "Protesters Take Control of Several Libyan Cities." February 19, 2011, http://www.thenews.com.pk/TodaysPrint-Detail.aspx?ID=31907&Cat=1.

Theidon, Cristina K. "Transitional Subjects: The Disarmament, Demobilization and Reintegration of Former Combatants in Colombia." *International Journal of Transitional Justice* 1, no. 1 (2007): 66–90.

Themnér, Lotta, and Peter Wallensteen. "Armed Conflicts, 1946–2011." *Journal of Peace Research* 49, no. 4 (2012): 565–575.

Thompson, William R. "Identifying Rivals and Rivalries in World Politics." *International Studies Quarterly* 45, no. 4 (2001): 557–586.

Thornton, Thomas. "Terror as a Weapon of Political Agitation." In *Internal War.* Edited by Harry Eckstein. Westport, CT: Greenwood Press, 1980.

Toft, Monica Duffy. *Securing the Peace: The Durable Settlement of Civil Wars.* Princeton, NJ: Princeton University Press, 2010.

Tripp, Charles. "The Sudanese War in International Relations." In *After the Cold War: Security and Democracy in Africa and Asia.* Edited by William Hale and Eberhard Kienle. London: Tauris Academic Studies, 1997.

Tzu, Sun. *The Art of War.*

Ukraine News One. "Putin Awards Russian TV for Crimea Coverage: Kremlin-Controlled Media Accused of Propaganda Role." May 6, 2014.

Universidad Nacional de Colombia. *Observatorio de Procesos de Desarme, Desmovilización y Reintegración, Dinámicas de las Autodefensas Unidas de Colombia* (Bogotá, August 2009.

US Army Doctrine and Training Publication (ATP) 3-05, *Unconventional Warfare.* Washington, DC: Headquarters, Department of the Army.

US Department of Homeland Security, Immigration and Customs Enforcement. "Mohamad Youssef Hammoud Sentenced to 30 years in Terrorism Financing Case." News release, January 27, 2011, http://www.ice.gov/news/releases/1101/110127charlotte.htm.

US Government. *Guide to the Analysis of an Insurgency 2012.*

US Government Interagency Counterinsurgency Initiative. *U.S. Government Counterinsurgency Guide.* Washington, DC: Bureau of Political-Military Affairs, Department of State, 2009.

Valentino, Benjamin A. "Why We Kill: The Political Science of Political Violence against Civilians." *Annual Review of Political Science* 17 (2014): 89–103.

Van Engeland Anisseh, M., and R. Rachael. *From Terrorism to Politics.* Vermont: Ashgate, 2007.

Van Essen, Jonte. "De Facto Regimes in International Law." *Utrecht Journal of International and European Law* 28, no. 74 (2012): 31–49.

Vargas, Jose Antonio. "Spring Awakening: How an Egyptian Revolution Began on Facebook." *New York Times*, February 17, 2012.

Vogt, Heidi. "Made in Africa: a Gadget Startup." *Wall Street Journal*, July 10, 2014.

Wahab, Bilal. "How Iraqi Oil Smuggling Greases Violence." *The Middle East Quarterly* 13, no. 4 (Fall 2006): 53–59.

Wallensteen, Peter, and Margareta Sollenberg. "Armed Conflicts, Conflict Termination and Peace Agreements, 1989–96." *Journal of Peace Research* 34, no. 3 (1997): 339–358.

Walt, Vivienne. "How Did Gaddafi Die? A Year Later, Unanswered Questions and Bad Blood." *Time*, October 18, 2012, http://world.time.com/2012/10/18/how-did-gaddafi-die-a-year-later-unanswered-questions-and-bad-blood/.

Waterman, Shaun. "Terrorists Discover Uses for Twitter." *Washington Times*, April 28, 2011.

Weber, Max. *On Charisma and Institution Building.* Chicago: University of Chicago Press, 1968.

Weinmann, Gabriel. "New Terrorism and New Media." The Wilson Center Research Series, vol. 2. Washington, DC: Commons Lab of the Woodrow Wilson International Center for Scholars, 2014.

Weinstein, Jeremy M. *Inside Rebellion the Politics of Insurgent Violence.* New York: Cambridge University Press, 2007.

Wiegand, Krista E. "Reformation of a Terrorist Group: Hezbollah as a Lebanese Political Party." *Studies in Conflict & Terrorism* 32, no. 8 (2009): 669–680.

Wight, Martin. "International Legitimacy." *International Relations* 4, no. 1 (1972): 1–28.

Wilkinson, Steven. *Votes and Violent: Electoral Competition and Ethnic Riots in India.* New York: Cambridge University Press, 2006.

Wilson, Scott. "In Politics, Hamas Gains in the West Bank." *Washington Post*, June 29, 2005, http://www.washingtonpost.com/wp-dyn/content/article/2005/06/28/AR2005062801368.html.

Wintrobe, Ronald. *Rational Extremism: The Political Economy of Radicalism.* Cambridge: Cambridge University Press, 2006.

Wittig, Timothy. *Understanding Terrorist Finance.* London: Palgrave Mac-Millan, 2011.

Wood, Reed M. "Rebel Capability and Strategic Violence against Civilians." *Journal of Peace Research* 47, no. 5 (2010): 601–614.

Worsnip, Patrick. "South Sudan Admitted to U.N. as 193rd Member." *Reuters*, July 14, 2011, http://uk.reuters.com/article/2011/07/14/uk-sudan-un-membership-idUKTRE76D3I120110714.

Wright, Lawrence. *The Looming Tower: Al-Qaeda and the Road to 9/11.* New York: Vintage Books, 2006.

Yackley, Ayla. "Turkey Outlaws Kurdish Political Party." *The Globe and Mail*, 2003.

Zanotti, Jim. "Hamas: Background and Issues for Congress." CRS Report no. R41514. Washington, DC: Congressional Research Service, December 2, 2010.

Zedong, Mao. *On Guerrilla Warfare.* Urbana: University of Illinois Press, 2000.

Zisser, Eyal. "Hizballah in Lebanon: At the Crossroads." *MERIA* 1, no. 3 (1997).

Zuckerman, Ethan. "The First Twitter Revolution?" *Foreign Policy*, January 14, 2011.

Zürcher, Christoph, Carrie Manning, Kristie D. Evenson, Rachel Hayman, Sarah Riese, and Nora Roehner. *Costly Democracy: Peacebuilding and Democratization after War.* Stanford: Stanford University Press, 2013.

INDEX

A

Administrative structure, and transition from resistance to governance, 170–173

Adversary critical resource assessment, 122–125

Afewerki, Isaias, 4, 171

Affection, as motivation for insurgent interest and identification, 61

Afghanistan
  civil war in, 165
  Dutch disease in, 136
  foreign fighters in, 107–108
  information technology use in, 90, 93
  opium produced in, 113
  phases of insurgency in, 8
  as potential case study, 208
  support for Crimean referendum in, 156
  Taliban in, 145, 164–166
  US attempts to disrupt financial bases of insurgencies in, 115

Africa, 208–209; *See also individual countries*

African National Congress (ANC), 37
  and Defiance Campaign of Unjust Laws, 41–42
  FRELIMO support for, 168
  types of violence used by, 79

African Union Mission to Somalia (AMISOM), 124

Agency slack, 196

Al-Banna, Hasan, 106

Al-Nusra, 111

Al-Qaeda
  and EIJ, 108
  information technology used by, 92
  roots of, 107–108
  threat financing for, 106–108
  types of violence used by, 80

Al-Qaeda in the Islamic Maghreb (AQIM), 111

Al-Shabaab
  tactical economic analysis of, 120–130
  threat financing by, 110, 111

Al-Zawahiri, Ayman, 108n.c

Amal, 34, 43

American Revolution, 4

AMISOM (African Union Mission to Somalia), 124

Analyze phase (TEA), 118–125

ANAPO (National Popular Alliance), 45

ANC; *See* African National Congress

Angolan War of Independence, 182–183

Anti-Fascist underground (Italy), 109

AQIM (al-Qaeda in the Islamic Maghreb), 111

Arab Spring, 190

Arenas, Jacobo, 113

Armed component, 25

Armed force, groups' development of, 150–152

Asia, 208; *See also individual countries*

Assassinations, 78

Asymmetric information, 195–196

AUC (Autodefensas Unidas de Colombia), 63–66

Auxiliary component, 25, 48

Aziz, King Saud bin Abdul, 106

Azzam, Abdullah Yusuf, 107

B

Balkan Ghosts (Kaplan), 58

Bangladesh, 37

Bank of Algiers, 102

Bank of England, 102

Bardot, Brigitte, 112

Baselines (analyze) phase (TEA), 118–125

Base of support
  creating, for legitimacy, 148–150

Index

Palestine, Hamas in, 163–164

Palestinian Legislative Council (PLC), 149, 164

Palestinian Liberation Organization (PLO), 80–81, 203

Palestinian National Authority (PNA), 37

Palestinian resistance movements, 101

Parallel financial systems, 106–108

People's Democracy Party (HADEP), 44

People's Front for Democracy and Justice (PFDJ), 148, 171

Perceptual territory, 28–29

Personal grievances, as motivation for insurgent interest and identification, 61

Peshawar Accord, 165

PFDJ (People's Front for Democracy and Justice), 148, 171

Phases of contemporary resistance, 3–22

  definitions of resistance and insurgency, 3–4

  existing phasing constructs, 10–13

  further study on, 15

  importance of understanding, 3

  literature on, 4–10

  and stages of organizational growth, 13–15

  values of variables related to, 16–22

Philippines, 101, 104, 105

PIRA; See Provisional Irish Republican Army

PKK; See Kurdistan Worker's Party

PLC (Palestinian Legislative Council), 149, 164

PLO (Palestinian Liberation Organization), 80–81, 203

PMESII-ASCOPE assessment, 118–122

PNA (Palestinian National Authority), 37

Polish underground, 103

Political party(-ies); See also Institution-based public interface; individual parties

  achieving legitimacy as, 152–153

  as fronts for insurgent groups, 38–41

  independent of insurgent groups, 44–45

  as legitimate face for illicit organizations, 31

  rebel groups subordinate to, 41–43

  use of violent groups by, 37

Political violence, studies of, 57–58

Popular Movement for the Liberation of Angola, 182–183

Post-Cold War era

  civil war dynamics in, 185–191

  funding in, 27

  nature of internal conflicts in, 181

Prabhakaran, 151

Principal–agent relationship, 38, 195–197

  political parties as fronts for insurgent groups, 38–41

  rebel groups subordinate to political parties, 41–43

Prioritize (find) phase (TEA), 125–127

Propaganda, 203, 205–206

Provisional Irish Republican Army (PIRA)

  internal characteristics of, 197

  and legitimacy vs. legality, 46

  and Sinn Féin, 39–40

  threat financing for, 102, 104

  use of political parties by, 25

Public component, 25–51

  and change in nature of warfare, 26–29

  defined, 25–26

  within existing doctrine, 48–51

  goods-based interface of, 32–36

  institution-based interface of, 37–45

  as "public interface," 29–32

  strategic value of, 45–47

253

www.ingramcontent.com/pod-product-compliance
Lightning Source LLC
Chambersburg PA
CBHW052110020426
42335CB00021B/2701